DEC 02

DATE DUE

MAY 21 03		
OCT 20 06		
NOV 20 06		
JAN 04 07		
NOV 10 08		
GAYLORD		PRINTED IN U.S.A.

weird nature

weird nature

John Downer

FIREFLY BOOKS

Published by Firefly Books Ltd., 2002

First Printing

National Library of Canada Cataloguing in Publication Data
Downer, John
       Weird nature
1st ed.
Includes index.
ISBN 1-55297-587-8 (bound).--ISBN 1-55297-586-X (pbk.)
       1. Animal behavior. I. Title.
QL751.D68 2002   591.5'1
C2001-903716-3

Publisher Cataloging-in-Publication Data (U.S.)
Downer, John.
  Weird nature / John Downer. -- 1st ed.
[168] p. : col. photos. ; cm.
Includes index.
Note: Companion book to the Discovery Channel series.
Summary: An illustrated look at the incredible and sometimes bizarre animal behavior from devious defenses to unique ways of moving.
ISBN 1-55297-587-8   ISBN 1-55297-586-X (pbk.)
1. Natural history—Miscellanea. 2. Nature. 3. Curiosities and wonders. I. Title.
508 21   CIP   QH48.D69 2002

Produced by BBC Worldwide Limited
80 Wood Lane, London W12 0TT

Published in Canada in 2002 by
Firefly Books Ltd.
3680 Victoria Park Avenue
Willowdale, Ontario M2H 3K1

Published in the United States in 2002 by
Firefly Books (U.S.) Inc.
P.O. Box 1338, Ellicott Station
Buffalo, New York 14205

Color separations by Radstock Reproductions
Printed and bound in Great Britain by Butler and Tanner Limited
Frome and London

contents

The mythical **unicorn**, tales of which were inspired by the discovery of the strange spiral horn of the narwhal whale.

introduction

Imagine a fantasy world where unicorns can materialize out of thin air, and where, on an ocean voyage, you might glimpse mermaids dancing in the breaking surf. Imagine, too, the sight of a terrifying devilfish — a flying fish so colossal that it blackens the sky with its delta-shaped body, and can bring down ships if it lands on deck. And imagine, if you dare, the fearsome giants that lurk beneath the waves — monsters with gigantic, suckered tentacles that can snatch a ship and pull it under.

Not so long ago, this strange world was viewed as reality, not fantasy. The field guides of the day, known as bestiaries, even chronicled the creatures' behavior, and travellers undertook fearful voyages to try and find them. The magical creatures seemed as real to people then as farmyard animals. Gradually, though, the natural world gave up its secrets, and the truth emerged. The monsters didn't exist, but behind the stories were real animals — not fantastically magical, but still strange and exotic in their own way.

The twisted horn of the unicorn, once proudly displayed in medieval banquet halls, was traced to a weird, one-horned whale called the narwhal; males use their horns to duel over females. Mermaids were demystified when the dugong, also known as the sea cow, was discovered. Its

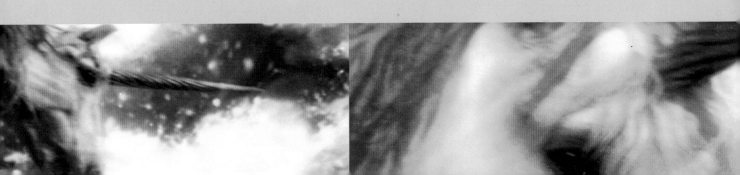

beauty is hardly the stuff of legend, but its human-like habit of bobbing upright in the waves, coupled with its fishy tail, may have been enough to fire the imagination of a bleary-eyed and homesick sailor. The animal behind the devilfish was probably a giant manta ray, a huge fish that occasionally launches itself from the sea and glides over the water. And the fearsome sea monster was a giant squid, an animal large enough to spawn many a tall tale, though perhaps not big enough to bring down a ship.

Bit by painstaking bit, science has revealed the truth of the natural world. In doing so it has dispelled many myths, but what it has revealed is no less extra-ordinary. This book, which accompanies a Discovery Channel series of the same name, explores some of science's most incredible discoveries about the behavior of real animals. Many of the revelations are just as fantastic as tales of mermaids or unicorns.

As we researched the themes of the series and book, it became clear that we were not only uncovering some of the strangest examples of animal behavior, we were also uncovering something about ourselves. In deciding what we regard as weird, we inadvertently define what makes us human.

Above: The sighting of a giant **manta ray** was probably the origin of the mythical flying devilfish.

When we compare ourselves to animals, we naturally find that the weirdest animals are often those that behave least like humans. But, paradoxically, animals also seem weird when they act like us. We think of ourselves as unique — a cut above the other species that inhabit the planet. When that illusion is shattered, we find the comparison strangely disconcerting; the usual reaction is a slightly embarrassed smile.

As I write this, I am sitting in Nairobi, Kenya, having just come back from filming elephants in the wild. Helping us on the trip was one of the world's leading elephant research scientists. When she interpreted the elephants' behavior for us, she talked in terms of the same emotions that underpin human behavior. It is the only way she can make sense of the complex actions she has observed over years of study. Likewise, primate scientists usually describe the behavior of their subjects in ways that echo human actions and motivations.

Until recently, one of the greatest crimes in science was to be accused of being anthropomorphic — that is, ascribing human thoughts and emotions to animals. As we learn more about the complexities of animal behavior, that position is beginning to appear untenable. We *are* animals, and the similarity between our behavior and that of the rest of the animal kingdom is more than mere coincidence.

In this book I have deliberately made human comparisons where I feel it is appropriate, and have used figures of speech that reflect human actions. I believe this not only makes animal behavior more readily comprehensible, but also helps us understand the animal background behind much of our lives.

It is only by understanding our deep links with the rest of the animal world that we can understand our place within it. Science may not have discovered unicorns, but by showing us the diversity of real animal behavior, it gives us a chance to reflect on our own. As we consider what defines *Weird Nature*, we in some way define ourselves.

John Downer

chapter one

The flying dragon, or **draco**. The "wings" of this spectacular lizard are membranes stretched between a series of elongated ribs that act as struts. The end result bears an uncanny resemblance to early designs for aircraft wings.

marvelous motion

If a visitor from an alien world examined the life of our planet, the strangest animal they would find would be humans. We differ from other animals in many odd ways, but physically our most idiosyncratic feature is the strange way we move. Biologically speaking, walking on two legs is very weird indeed. In apparent recognition of the fact, much of our technological development has tried to improve the way we get around. We build boats that sail the seas, submarines that explore the ocean depths, a host of wheeled vehicles that take us over land, and a plethora of planes, gliders, and other machines that traverse the skies. These vehicles carry us into worlds that were once the sole realm of other creatures — and, far from coincidentally, the vehicles' designs are frequently inspired by nature.

Although we marvel at the perfection in evolutionary design that allows birds to fly, or fish to swim, such natural motion is so familiar that it hardly seems strange at all. Only the less familiar ways of moving still tweak our weirdness detectors. Predictably, many of these are found in alien environments, such as the ocean.

Life began in the sea, and the sea soon became an experimental workshop that gave rise to many strange ways of moving. Because water supports the weight of an animal, the first body forms

were far simpler than those that later developed on land. Many of these early designs were so successful that they still grace the oceans today. One of the oddest is the aptly named Spanish dancer.

The Spanish dancer acquired its exotic name from its startling appearance and graceful movement. Like a flamenco performer, it has a

Right: The **flame scallop** claps its castanets to dance away from predators.

*Like a flamenco performer…with ruffles running along its body…the **Spanish dancer** shimmies through the tropical oceans.*

Left: The **Spanish dancer** moves through the tropical oceans by undulating its luscious red membranes. It is actually a sea slug.

luscious red costume, with ruffles running along its body and a gathering of red frills at the head. The frills are gills and play no part in movement, but the ruffles are fleshy extensions of the body; as they undulate, the Spanish dancer shimmies through the tropical oceans. Despite the romantic name, this beautiful creature is actually a sea slug.

…the **flame scallop** *…claps its shells together and darts through the water, like a pair of castanets with a life of their own.*

The rhythmic clapping of castanets accompanies every flamenco dancer and, appropriately, the coral oceans possess a living equivalent — the flame scallop. When danger threatens, this scallop claps its shells together and darts through the water, like a pair of castanets with a life of their own. Just a whiff of starfish essence is enough to trigger this comical escape mechanism. The scallop's

Below: The luminous bodies of **jellyfish** glow eerily as they swim through a wreck using gentle pulsations of their umbrellas.

secret is a primitive form of jet propulsion — as the shells close and clap together they squirt water out of two small openings on either side of the hinge. The resulting jets propel the shellfish smartly along.

Many sea creatures use some type of jet propulsion. Jellyfish take in and expel water by gently pulsating their umbrellas, but this crude design allows little

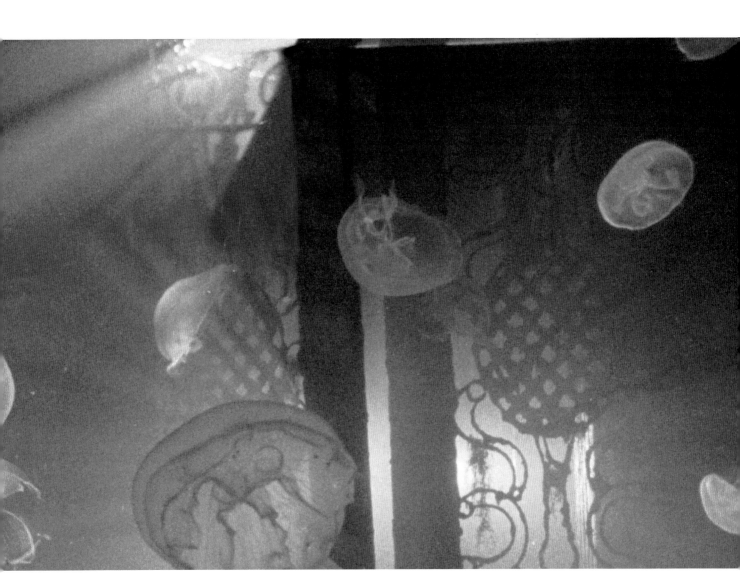

Opposite: The **nautilus** has perfected jet propulsion, expelling water through a moveable tube to propel it through the water.

Below: The **seahorse**'s armor prevents its body from flexing like ordinary fish; instead, its back fin moves it through the water.

directional control, and they often get stranded on beaches. Octopuses, squid, and cuttlefish have a more sophisticated version of jet propulsion. They expel water through a moveable tube called a siphon, which combines power with directional control.

The nautilus — a modern-day relative of the spiral-shelled ammonites familiar to fossil-collectors — also has a siphon, but it has another special feature. Its beautiful shell not only protects the soft-bodied animal within but also doubles as a buoyancy aid. The body of the mollusc inhabits the very last of a spiralling series of chambers inside the shell. By filling the inner chambers with a mixture of air and water, the nautilus achieves perfect buoyancy, allowing it to rise effortlessly during its nightly migration from the depths of the Pacific Ocean to the surface.

Of all sea creatures, fish have evolved the most sophisticated method of movement. They swim by undulating their bodies in an s-shaped movement and achieve extra thrust by flicking their tails and fine control by moving their fins. The weirdest fish are those that deviate from this basic technique.

The seahorse's strange, equine features come from a body covering of bony plates. This defensive armor prevents its body flexing like a conventional fish, but even so, it appears to glide magically though the water without apparent effort. Its secret is a transparent back fin that beats 20–30

17

times per second — so fast that it almost vanishes in a blur. The high-frequency shimmer propels the sea horse either forward or backwards. To change direction the sea horse uses a second set of fins, positioned just behind its equine head. Depending on which of these pectoral fins vibrates, the sea horse turns in the opposite direction. Because the sea horse is such a weak swimmer, it relies on camouflage to hide from predators and a prehensile tail to anchor it to seaweed.

The evolution of jointed legs gave the crustaceans — crabs, lobsters, and their relatives — a new means of getting around. The prize for weirdness goes to the spiny lobsters of Florida. It is not the movement of a single lobster that raises an eyebrow but the fact that, periodically, these animals join together for a mass conga across the sandy ocean floor.

Cued by the first autumn storms, they leave their rock crevices and assemble for the performance in lines of 50 or more. The leader sets the course and speed, while others, with their 10 legs kicking out sideways, join the conga line from behind. Like drunken party-goers, but using

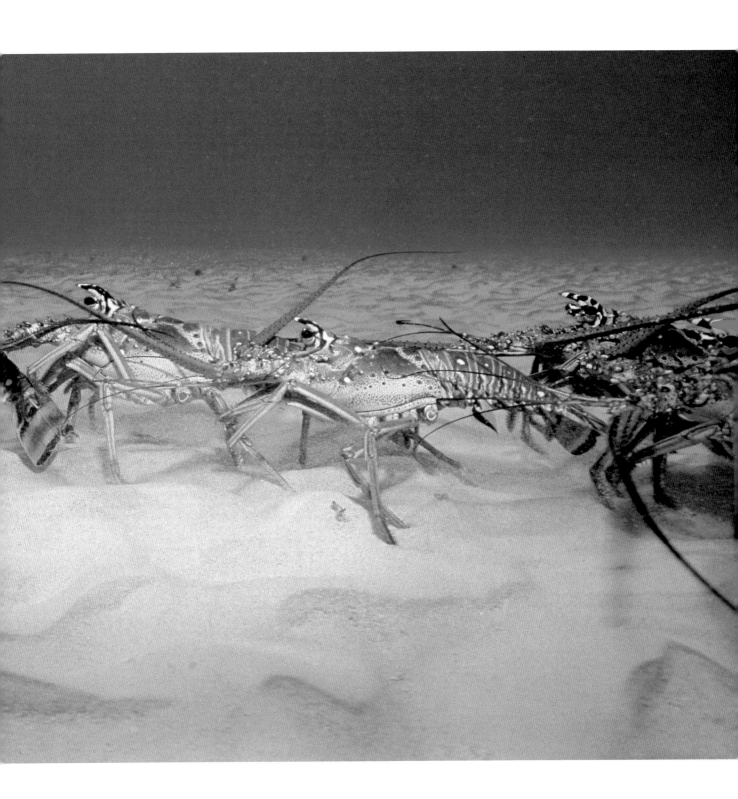

The **spiny lobsters** of Florida perform a mass conga towards the shelter of deeper waters.

*...the **lobsters** can quickly form a defensive circle like a Wild West wagon train.*

antennae rather than arms, they try to keep contact with one another as they swagger from the shallows to deeper water. Here they are protected from winter storms, and the cold slows down their bodily processes during the leaner months. Travelling in line reduces drag, and if predators strike, the lobsters can quickly form a defensive circle, like a Wild West wagon train.

The idea of legs really caught on among land animals. Without water to hold them up, land animals needed extra support, and jointed limbs attached to a skeleton became the best solution to moving heavy bodies around. But dragging a huge body along the ground was still a problem for many early land animals. Today the crocodile demonstrates one of the more unusual solutions. Usually the crocodile walks along in a belly crawl with its legs splayed out like a lizard, but because its underside drags along the ground it wastes a lot of energy. However, the crocodile can employ a neat contortionist trick: It can turn its feet 90 degrees so they face forward instead of out to the side. In this configuration, the crocodile can pull its legs close to its body to act as supports. This creates an odd stiff-legged gait — almost as if it were walking on stilts.

As well as walking oddly, Australian freshwater crocodiles adopt an equally strange method of moving at speed. The familiar galloping of horses, dogs, and other mammals is achieved by the limbs making contact with the ground in a sequence. This happens so quickly that it is very difficult to see —

Above: Life's a drag for a **crocodile**, unless it adopts this strange, stilt-like walk.

before the invention of photography, paintings and engravings of galloping horses showed the front or rear pair of legs touching the ground together. The freshwater crocodile is one of the few animals that really gallops that way and the result appears very strange indeed. When frightened, it pushes its rear legs back and down together while reaching out with its front legs. The front legs then push back as the rear ones swing forward to meet them. With both pairs of legs in synchronous but opposite motion, this crocodile can accelerate to 15 mph (25 km/h). It gallops so fast that it often becomes airborne, clearing any obstacles in its way. The freshwater crocodile gallops only to escape from danger, but larger, prehistoric crocodiles once used the same strange gait to outrun prey.

this **crocodile**...*gallops so fast that it often becomes airborne, clearing any obstacles in its way.*

Compared to anything in the natural world, the wheel seems one of the weirdest devices we use to get around. Invented about 5500 years ago, it was, until recently, thought to be a uniquely human technology. But bacteria have now been discovered that have a primitive type of spinning propeller, complete with axle, and several animals have been found that turn their whole bodies into wheels. One of these — the mother-of-pearl caterpillar — is surprisingly common, showing that even familiar creatures can still surprise us.

Above: The **mother-of-pearl caterpillar** escapes from predators by forming the shape of a wheel and rolling away at high speed.

*If it meets a predator, this **caterpillar** anchors its rear to the ground, recoils rapidly, and then rolls away backwards like a bright-green tire.*

Caterpillars usually progress by lifting each leg in turn to create a wave of grip that moves along the body. This makes them very stable but limits their speed to about half an inch (1 cm) per second. The mother-of-pearl caterpillar, however, can step on the accelerator. If it meets a predator, it anchors its rear to the ground, recoils rapidly, and then rolls away backwards like a bright-green tire. Mouth to tail, it completes around half a dozen revolutions during its escape. By turning into a wheel, the caterpillar moves some 40 times faster than its normal walking pace.

Below: The **Mount Lyell salamander** "transforms" itself into a rubbery tire for a fast getaway.

On the steep slopes of California's Sierra Nevada mountain range, the Mount Lyell salamander achieves a similar rock 'n' roll lifestyle. When disturbed, or when it needs to descend, the amphibian curls its head under its back legs,

wraps its tail along its body, and tucks its legs in. It not only looks like a black tire, it behaves like one. Rolling over and over, it bowls down any slope with ease, its rubbery body absorbing the impact of bounces. When it reaches safety or runs out of slope, it uncurls and crawls to shelter. Even the salamander's normal motion is peculiar — it curls the tip of its tail against the ground to create a fifth limb, giving extra purchase on steeper slopes.

Below: The **Mexican jumping bean** is a novelty toy because of its weird leaping. The secret is a caterpillar inside.

Moving unexpectedly is a good way to surprise predators. In the Alamos desert of Mexico some plant seeds do just that. Known as Mexican jumping beans, their jerky movements are so intriguing that millions are exported to toy stores each year. The secret of the baffling motion is a moth caterpillar hidden inside.

The cycle starts when a moth lays her eggs in the flower of the jumping bean shrub. As the three-seeded pod develops, the caterpillars hatch and eat their way into the seeds. When the pod matures, the seeds separate and fall. Those containing caterpillars soon start to jump.

To perform its trick, the caterpillar first weaves a silken web against the seed wall. Then, by grasping this web with its forelegs and jerking violently, the caterpillar transfers the full force of its movement to the capsule. The sudden jolt startles any animal that tries to eat the seed. By jerking repeatedly, the caterpillar can even roll the seed into the shade to find shelter from the sun's heat.

The jumping bean's acrobatics last for several months. The caterpillar thencuts a circular trapdoor in the seed wall and forms a cocoon. A few weeks later it escapes as an adult moth. An adult female then has only a few days of life left to mate, find a shrub, and lay eggs to complete the life cycle — providing she has not ended her days in a toy store.

Because jumping provides such a good means of escape, many familiar animals, such as frogs and kangaroos, are experts at it. Less well-known are the jumping talents of one of our distant relatives.

The bushbaby is a primitive primate that has the extraordinary ability to bounce around like a living rubber ball. A single leap can carry it 7 feet (2 m) into the air, or six times its body length — equivalent to a human high jumper clearing two stacked double-decker buses. This feat needs more than sheer muscle power — it actually requires a clever power-amplification system. As the bushbaby lands, it stores energy in its tendons, which act like catapults. When it

Overleaf: A multi-flash picture showing the arc described by a leaping **bushbaby**. For the first half of the leap the bushbaby stays vertical. Then, at the apex of the leap, it starts to reach out in the direction it falls.

Opposite: A **bushbaby**'s legs absorb the shock of impact and convert it into energy to propel the next leap.

Below: A group of **sifaka lemurs** pogo across the forest floor.

takes off again, the power is transmitted down the leg and helps extend the ankle and the foot — literally giving the bushbaby a spring in its step. Predators find it almost impossible to catch.

The sifaka lemurs of Madagascar show a similar talent for jumping, but they progress in a series of vertical hops that make them appear to pogo across the forest floor. To add to their absurd appearance they hold their hands above their heads for balance as they bounce to the next tree.

*The **Wallace's tree frog**…can glide as far forward as it drops vertically. It can even maneuver in flight….*

Frogs are rightly famed for their leaping prowess. Some species push the envelope by making apparently suicidal leaps from the tops of trees. To slow their descent they use a kind of parachute. They have enlarged hands and feet with wing-like webbing, and some also have skin flaps on the margins of their arms and legs. Frogs that have only a small amount of webbing are known as "ploppers" for the simple reason that they drop almost vertically in a slow free fall. Other species, such as the Wallace's tree frog of Malaysia, can glide. The Wallace's tree frog has such extensive webbing between its toes that it can glide as far forward as it drops vertically. It can even maneuver in flight to aim for broad-leaved plants that help absorb the impact of landing.

Some lizards have also joined the flying club. The flying gecko from the forests of Southeast Asia has webbed feet, a fringed tail, and flaps of skin along its flattened body that spread out like wings. But the ultimate design belongs to a lizard that lives in the same forests — the flying dragon, or draco. The "wings" of this spectacular lizard are made from membranes stretched between a series of specially elongated ribs that act as struts. The end result bears an uncanny resemblance to early designs for aircraft wings, but the lizard's version is far more sophisticated since its wings can be opened and closed at will. The lizard's flattened body improves its aerodynamics, and its slender tail acts as a counterbalance. On average, each flight carries it roughly 25 feet (8 m).

Opposite: The **Wallace's tree frog**'s extensive foot webbing allows it to glide long distances.

Overleaf: These multi-flash time studies show the glide angles of various **frogs**. Those with more webbing on their feet glide a farther distance.

Various small squirrels, possums, and flying lemurs have also developed skin flaps that help them glide between trees, but perhaps the weirdest gliders of all are the flying fish.

Flying fish leave the water to escape from aquatic predators, though this strategy can make them vulnerable to predatory frigate birds. Of the 40 species that soar over the world's warm oceans, one of the largest is the California flying fish, which grows to an enormous 1.6 feet (0.5 m) in length. All flying fish owe their aeronautical skills to wing-like pectoral fins and, in some cases, expanded pelvic fins as well. To become airborne, a flying fish accelerates towards the water surface with fins folded against its torpedo-shaped body. As it rockets through the surface the still-submerged tail whisks up additional thrust and propels the fish to a speed of approximately 19 mph (30 km/h). The fins then open and the fish glides about 1 yard (1 m) above the surface at a cruising speed of around 10 mph (16 km/h). On a single glide they may travel as far as 591 feet (180 m) and reach a height of 33 feet (10 m). Some fish can even make several consecutive glides without submerging, by using a flick of the tail to propel themselves back into the air. On these multiple glides a flying fish may remain airborne for 43 seconds and travel more than 1300 feet (400 m).

The least likely flying animal must be the snake — it seems particularly handicapped as it has neither fins to expand nor limbs to act as rigging for webbing or wings. Even so, the golden tree snake of Southeast Asia, among others, makes a decent stab at the art of gliding. Like most gliders, it uses a tree to gain height. Before takeoff, the snake makes a loop with the front part of its body and then lowers this loop into the open space below. Then the snake springs forward, pushing its tail against a branch and flattening out its body until it has doubled in width. The ribbon-like body also becomes slightly concave to act like a wing.

Opposite: The **flying gecko** has every edge of its body expanded into a membrane that acts as a wing as it glides through the forest.

Opposite: A multi-flash time study of a **golden tree snake** launching itself into the air. It forms a loop for maximum thrust.

As the snake glides, large s-shaped waves pass along its body as though it were swimming. It controls direction by banking its body like a turning airplane. To make a sharp 90-degree turn, it simply flips its tail. A single flight may carry the snake 30 feet (10 m) or more.

To overcome their lack of limbs, snakes have evolved other strange forms of locomotion. One of the oddest occurs in deserts. Loose sand grains are a particular problem for snakes, as their normal locomotion relies on having a firm surface to give purchase. The North American horned rattlesnake solves the problem by forming its body into moving loops that make contact with the ground in only two places at a time. As these contact points shift from front to rear, the snake "corkscrews" sideways over the sand. The technique is so successful that the sidewinder of Southern Africa's Namib desert independently evolved an identical way of travelling.

Our North American deserts are home to another curious sand-dwelling snake, but this species burrows beneath the sand instead of spiralling across it. Sand behaves more like a liquid than a solid, and as the name suggests, the sandswimmer's solution is to swim through it.

The sandswimmer's wedge-shaped snout parts the sand like the bow of a boat breaking through water, and the sand grains slide over the body's ultra-smooth scales like water off a duck's back. Using s-shaped swimming

*Sand behaves more like a liquid than a solid, and as the name suggests, the **sandswimmer**'s solution is to swim through it.*

movements, the snake slides through the sand almost effortlessly. Special valves in the snake's nose stop sand from blocking its airways and suffocating it.

Among mammals there is one method of travelling that is especially rare — walking on two legs. Our unusual gait places humans among the weirdest animals on the planet. The only other mammals that usually move on two legs are kangaroos and kangaroo rats, but since they jump rather than walk, they are hardly in our league. But we do share our strange gait with birds.

All birds are bipedal but, as most of them fly, the fact tends to go unnoticed. It is only flightless birds, such as the ostrich, emu, and kiwi, that are obvious two-legged exponents. To us, these flightless birds seem slightly comical — we see ourselves reflected in the strange way they move around. Penguins, which stand upright like ourselves, provoke an even stronger reaction. But there is also a flying bird that appeals to our sense of the absurd.

Above: The **side-winding rattlesnake** avoids that sinking feeling by corkscrewing across the sand. A wave of contact points passes along its s-shaped body.

The roadrunner found fame around the world in a certain familiar, long-running cartoon series. The animal that lies behind the screen legend is no less entertaining in real life. It is a kind of cuckoo that lives in the scrubby deserts of the southwestern U.S. Although it can fly, it prefers to run, chasing down prey in a comical sprint. It is the fastest runner out of all the flying birds, reaching speeds of 26 mph (42 km/h). Not only is it swift, it can also turn abruptly at right angles without slowing down. The key to its maneuverability is its tail, which acts as a rudder. The tail also doubles as a brake, flipping up when the bird needs to slow down.

Below: The real-life **roadrunner**, whose comical sprint inspired the cartoon character.

To gain energy for its mad maneuvers the

roadrunner *has a dark patch of skin on*

its back that serves as a solar panel.

To gain energy for its mad maneuvers
the roadrunner has a dark patch of skin on its
back that serves as a solar panel. At sunrise it
erects its feathers to expose this patch to the
sun's rays. When the bird is up to
temperature, it speeds after lizards or snakes,
which it spots with its keen eyesight.

The ancestors of birds were the
dinosaurs, and many of these were also
bipedal. Among the most famous was the
predatory *Tyrannosaurus rex*. Today miniature
equivalents still hunt down their prey in the
same way.

In the deserts of Mexico and the
southwestern U.S., the collared lizard
behaves just like a pygmy tyrannosaur as it
chases smaller lizards for food. It starts the
chase on four legs but soon shifts to its hind
legs for extra speed — bipedal running

gives it the edge. Only a handful of lizards can move on two legs, and most are desert dwellers because the open terrain makes it easier for them to get up on their back legs and maintain their speed. Other bipedal racers include the frilled lizard of Australia, which runs comically on two legs when chased by predators, and the basilisk of Central America. The basilisk can even run bipedally on the surface of water — a miraculous feat that gave rise to its alternative name: the Jesus Christ lizard.

If two-legged walking is so exceptional, then how did our strange locomotion arise? Our closest relatives provide the clue.

In the mangroves of Borneo the proboscis monkey often walks on two legs, hands held above it head, as it wades through the water. Lowland gorillas and bonobos (pygmy chimpanzees) occasionally walk in a similar way when they cross flooded plains. Some scientists believe we had an aquatic period in our evolution and the need to keep our head above water gave rise to our bipedal gait. A more convincing explanation springs to mind when we watch apes and monkeys carrying fruit or other provisions — walking on two legs frees the hands for grasping. When early hominids became hunter-gatherers, the need to carry supplies became increasingly important, and walking on two legs was the inevitable outcome.

Whatever the reason, our strange way of walking is exceptional in mammals and rare in the rest of the animal world. In terms of locomotion, it is *we* who are weird. And the same can be said of the way we reproduce.

Opposite: Resembling a miniature tyrannosaur, the **collared lizard** can run faster on two legs than four.

41

chapter two

In a courtship embrace: the female **Siamese fighting fish** will lay her eggs in the nest of
air bubbles blown by her partner.

bizarre
breeding

In nature's weirdness stakes, our way of breeding must be a clear front-runner. Few animals have such a prolonged courtship or spend so much time bringing up their young. Few animals are as sexually active as we are, yet, ironically few have such low fertility. Most animals have sex only when they are fertile, which, unlike in our case, usually happens in seasons. Most animals have only perfunctory orgasms, and these rarely occur in both sexes. In nature, childbirth is almost always painless and mostly involves more than one offspring. Most animals simply die when they pass breeding age and so do not experience menopause. In fact, we are so unusual that it seems almost perverse to consider the breeding practices of other animals as strange. Perhaps it is not surprising, therefore, that some of the most unusual examples of animal courtship and sex are those that remind us of ourselves.

Our practice of offering gifts in courtship, for example, is far from unique; the bowerbirds of Australia and New Guinea also have an eye for the kind of presents their partners adore. These pigeon-sized birds use twigs, leaves, and moss to construct an elaborate structure, known as a bower, on the forest floor. The bowers are more like bachelor pads than nests, designed by males to attract and seduce a mate. To add to the bower's appeal the male indulges in home decorating, tastefully adding ornaments

The **satin bowerbird** entices a female into his bower, which he has decorated with blue odds and ends.

of feathers, pebbles, berries or shells. Bowerbirds even steal human trinkets, such as shiny coins, pen caps, bits of pottery, keys, or jewelry, to help create the perfect romantic mood. As the male waits for a female to call, he passes his time tinkering with home improvements and rearranging the decorations. Then he struts and sings on the doorstep to persuade a female to enter the bower and mate. When a female succumbs, she promptly builds a nest close by. The male, meanwhile, stays in his bower and tries to persuade other females to join his romantic interlude.

The male uses its bill as a paintbrush to paint the walls of his u-shaped bower with chewed berries. His favorite color scheme is blue....

Each of the 17 species of bowerbird has a distinctive bower design and decorating scheme. As a suburban touch, some even plant lawns of moss around their bower. The showier species tend to spend less effort on their bowers, while drab ones build flashy bowers to compensate for their lack of sex appeal.

The satin bowerbird, which is named for its iridescent plumage, takes home decorating further. The male uses its bill as a paintbrush to paint the walls of his u-shaped bower with chewed berries. His favorite color scheme is blue, and he avidly collects anything that color, including feathers, berries, and flowers. He finds many of these knickknacks in the forest, but he is equally happy to steal ornaments from his neighbors' bowers. Young males often fall victim to such petty thieving, and consequently fail to accumulate sufficient wealth to impress a female. Experienced males, however, may mate with dozens of different birds in a season.

If the bowerbirds' behavior seems strange, it is because of the parallels with our own behavior. The same applies to dancing. Just like us, some birds go "clubbing" to attract a mate. In an effort to impress, the male manikin stages the most extraordinary solo dances. The most sensational is performed by the red-capped manikin, who does a more than passable impression of Michael Jackson's "moondance."

Above: **Scorpions** perform an elaborate courtship dance.

Another unlikely dancer is the scorpion. In courtship the male and female link claws and perform a promenade. As they dance, the male drops a package of sperm and waltzes his partner over it until she sucks up the present into her genital pore. The ungrateful female sometimes ends the dance by eating her partner, so the male may take the precaution of drugging her first with a sting.

Even the king of the beasts has a claim to weird sexual behavior. The lion's speciality is endurance — a courting couple's sexual marathon can last for several days. During this period lions mate once every 25 minutes and notch up an impressive 300 copulations. Why lions are so passionate is unknown, but the lioness seems to need to mate this many times to stimulate ovulation.

At first the female seems to find the process painful, and at the point of climax she often lashes out at the male. His reaction is to bite the scruff of her neck, using a scruff-hold otherwise reserved for wayward cubs. The lioness is changeable, and her angry reluctance quickly turns to enjoyment. Soon she begins to take the lead, slinking seductively around the increasingly reluctant male. The male valiantly tries to oblige, but when fatigue finally sets in the lioness may have to use her paw to shake him from post-coital slumber.

Lions may appear devotees of tantric sex, but the actual act is very short — copulation lasts just 21 seconds on average. Perhaps a better contender for the title of nature's Casanova is an Australian marsupial mouse known as the Antechinus; the male Antechinus has the dubious distinction of literally mating himself to death.

Opposite: Male **antechinus** have the dubious distinction of mating themselves to death.

Below: **Lions** perform a sexual marathon that might involve 300 copulations.

Once spring arrives, the thoughts of the male Antechinus turn to sex and little else. These tree-dwelling sex maniacs indulge in bouts of lovemaking that may last 12 hours at a time. As one passionate session ends, another begins with a different female. The orgy may involve 16 partners, and the male that services all these females has little time left to eat, drink, or sleep.

…many plummet to the ground in a state of exhaustion. Some even expire while in flagrante seducto. In just two weeks all the males are dead, and the females are left behind to bring up the babies.

After many nights of wild passion the males begin to look a little worse for wear — they become thin and haggard, and their hair starts to fall out. To make matters worse, battles with rivals inflict further injuries. Soon the tree-top orgy becomes risky, and many plummet to the ground in a state of exhaustion. Some even expire while *in flagrante seducto*. In just two weeks all the males are dead, and the females are left behind to bring up the babies.

Although the sexual activity of these animals appears bizarre, they are actually just indulging in an extreme variation of straight sex. In the ocean things are not so straightforward — the underwater world is a hotbed of sexual confusion and gender-bending.

Basslets are small coral fish that swirl around in large shoals of mostly females. The few males that shimmy through the shoal stand out from the drabber females because of their bright-pink costumes. But the females' sexuality is flexible — if a male is lost or killed, a female can perform a quick sex change to take his place. "She" becomes "he" by donning more colorful attire and developing male sex organs. No matter how many males are removed, there are always females ready to replace them. Although these transsexuals function as males, they never quite lose their maternal instincts, and they will still make nests and guard their young, unlike true males.

In clownfish, the sex change occurs in the opposite direction. These fish live among the stinging tentacles of sea anemones in groups made up of a male, a female, and several juveniles. If the female dies, then the male's sexual organs stop working, and previously dormant ovarian cells inside his body start making egg cells. He then takes the female's place. To fill the vacuum left by this sex change, one of the juveniles then turns into an adult male. In this way, the sexual make-up of the group stays constant, and no breeding time is wasted.

Opposite: Gender-bending on the coral reef. A male **clownfish** will change into a female if the latter dies.

Below: A female **basslet** will transform into a more colorful male if the resident male is lost from the shoal.

…some deep sea fish…are simultaneous hermaphrodites, possessing both working male and female sex organs.

More than one hundred species of fish are this kind of sequential hermaphrodite, starting life as one sex and changing into another. But there are some deep-sea fish which are simultaneous hermaphrodites, possessing both working male and female sex organs. During courtship these fish first act as males and release sperm; then they change roles and release eggs, just like females. This sexual switching helps fish that have difficulty finding mates in the vastness of their deep-sea home. Although such behavior is unusual in fish, among molluscs it is normal.

The sea hare is one of the more extraordinary molluscan hermaphrodites. Named after a front pair of tentacles that look like a rabbit's ears, the sea hare actually bears a closer resemblance to a slug. When two sea hares meet, one of them behaves like a male, extending a tentacle-like penis and penetrating the genital pouch of the other. After a few minutes they reverse roles. Others often join in the orgy, forming a chain of mating sea hares; the leader acts as a female, the one at the back acts as a male and those in between are simultaneously female and male. Sometimes the leader turns and joins the back of the chain to create a revolving wheel of hermaphrodite sea hares. The result of all this activity is an astonishing fecundity — a single sea hare can deposit more than 180 million eggs at a time, in a tight, winding mass of green ribbons.

For mass sex, no creatures match the reproductive antics of a coral reef. A coral reef consists mostly of dead deposits of the skeletons of past generations — only the outer layer is alive. The living coral organisms, or polyps, are permanently attached to their stony skeletons and, like many sea creatures, can have sex only at a distance — their eggs and sperm are simply released into the sea and left to their own devices. Such a haphazard system requires timing to succeed, which is why many corals synchronize the release of their sex cells. A combination of sea temperature and lunar phases provides the cue for such "mass spawnings," when billions of polyps release their sex cells at precisely the same moment.

The biggest sexual event in the natural world happens when the corals of the Great Barrier Reef spawn. For 1240 miles (2000 km) along the northeast coast of Australia, the sea turns into a multicolored, reproductive soup for just a single night. Predators are unable to devour such a mass of food, allowing most of the corals' offspring to survive.

Many animals opt for less wasteful ways of ensuring that some of their offspring survive — they produce fewer offspring, but they provide some kind of childcare.

"I'm forever blowing bubbles" is a refrain particularly apt for those that, like Siamese fighting fish, care for their young by literally putting them in a bubble. These fish are famed for their aggression — in Thailand people bet on the outcome of fights between them. But the males have a tender side, too. They prepare a home for their future young by taking gulps of air and blowing saliva-coated bubbles, which collect at the water's surface as a glistening froth. When a female arrives she blushes to display her enthusiasm, a pattern of diagonal stripes appearing on her skin. She then swims under the bubble-nest, where the male embraces her and fertilizes her eggs. Then, picking up the eggs in his mouth, he spits them, one at a time, into his bubbles.

The male's mood then changes and he aggressively drives the mother away, but his role as father continues as he conscientiously watches over the developing eggs. Any that fall are carefully retrieved and spat back into a bubble. The eggs hatch within two days, and the fry hang upside down from the bubble-nest, feeding from their egg sacs. Their attentive father continues to save any that fall from the nest and only abandons his parental duties when the fry finally swim free.

Several tropical frogs, known as foam-nesters, also build a nest of bubbles. The mother exudes a fluid and beats it into microscopic bubbles with her hind legs. She then lays her eggs inside, and her mate, who has clung to her back throughout, fertilizes them. As the parents leave, the outer bubbles harden to form a protective case that encloses a foamy core of several thousand eggs. This foam nursery provides shelter from predators, bacteria, and sunlight, as well as preventing dehydration. Because the foam is mostly air it supplies all the embryos' oxygen needs until well after hatching. The nest then disintegrates,

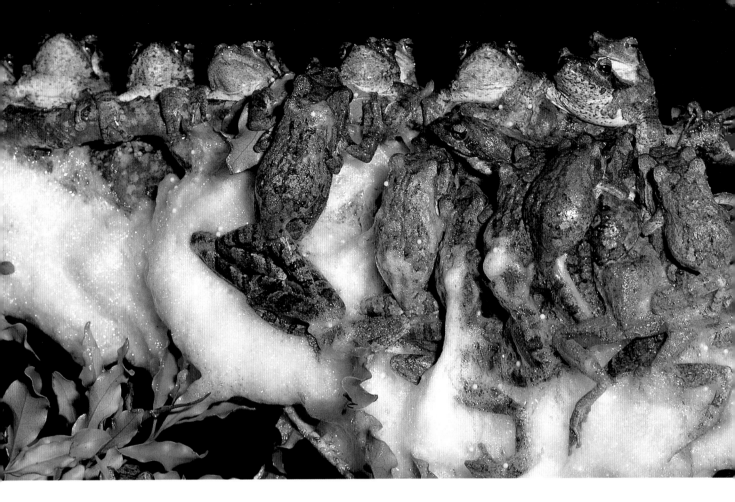

Above: **Foam-nesting frogs** create a nest of bubbles to house and protect their eggs. When the eggs hatch, the nest disintegrates and the tadpoles drop into the water below.

and the young emerge from their crowded apartment and, all being well, drop into water below.

Bubble-nesters make a safe haven for their young but then leave them to their own devices. Some creatures take the role of parenting more seriously, and some even seem to make unnecessary work for themselves. The Amazon splash tetra has the most labor-intensive childcare arrangements of any fish.

The expression "like a fish out of water" perfectly describes the bizarre breeding habits of these little fish. At the side of the river, males congregate in rows beneath overhanging leaves to stage a leaping contest. As if challenging their piscine nature, the contestants leap out of the water to find a place for their mates to lay eggs. Each male then lines up side by side with his partner, beneath the chosen site, and gives the ultimate performance in synchronized swimming.

Opposite: The astonishing synchronized leaps of the **splash tetra**, a fish that has the odd habit of laying its eggs out of the water.

In a movement too fast for the human eye to see, the female cues the male by nudging him with her head. Then they rocket from the water in perfect unison, so close together that they appear to be a single fish. Their landing is almost as miraculous, for they even stick to vertical leaves. In an instant the female lays an egg, the male fertilizes it, and they tumble back into the water. The synchronized leaping continues until all the female's eggs have been laid on many different leaves. But the father's work has only just begun. The eggs need to be kept constantly wet, so he remains continually occupied until they hatch, splashing them with well-aimed flicks of his tail.

…they rocket from the water in perfect unison, so close together that they appear to be a single fish.

The reason for such labor-intensive childcare is that the eggs are protected from underwater predators. But in the weird world of fish, not everything that gobbles eggs is necessarily a predator — some fish actually suck up their own eggs or young and rear them in their mouths.

Mouthbrooding is a rare and specialized form of parental care that reaches its zenith in the cichlids of South America and Africa. The mother picks up her eggs soon after laying, and they remain in her mouth until they hatch. Here they are protected and kept well-supplied with oxygen as they roll around. Mouthbrooding often continues well after hatching — the fry may leave the mouth to forage, but they rush back inside at the slightest sign of danger. It is usually the mother that mouthbroods, but parental duties are sometimes shared, or the role may be taken solely by the father. In one cichlid species from Africa's Lake Tanganyika, the fry move into their father's mouth when they have outgrown that of the smaller female. Remarkably, the mother has to find the original father among large shoals of very similar fish. She needs to choose well — an unrelated male would be better off swallowing rather than incubating the young.

Like a cuckoo, the mochokid catfish exploits the fact that mistakes can go unnoticed. The catfish spawns alongside a spawning cichlid, ensuring that the unsuspecting fish accidentally picks up the wrong eggs along with her own. The catfish eggs develop more quickly than those of their surrogate mother, and, when they hatch, the fry consume their host's young. Oblivious to the carnage inside her mouth, the cichlid raises the young catfish as if they were her own.

Opposite: The young of the **Surinam toad** hatch under the membranous skin of the female and break free when grown into tadpoles.

Below: The male **Darwin's frog** rears his young inside his vocal sacs.

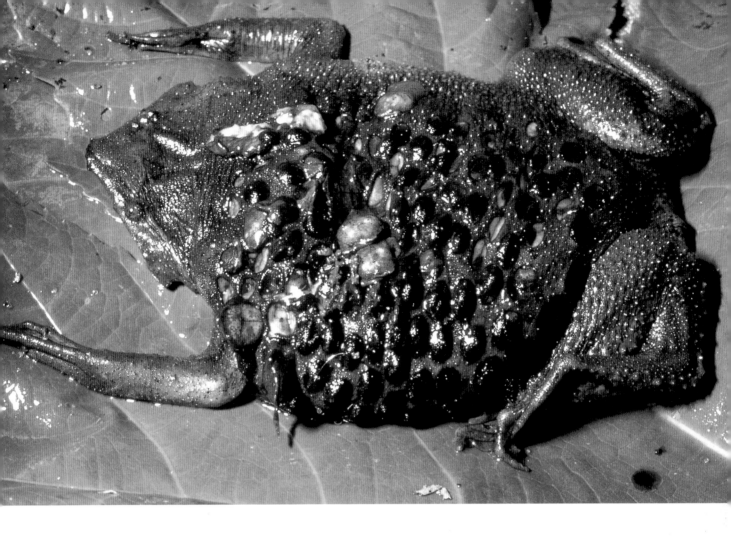

*...**Darwin's frog** gives literal meaning to the phrase "having a frog in the throat."*

Similarly, Darwin's frog gives literal meaning to the phrase "having a frog in the throat." The male guards the eggs until they hatch. He then takes them up in his mouth and rears them inside his vocal sacs until they eventually hop out.

Not to be outdone, the rare "gastric frog" of Australia rears her young in her stomach. During brooding, the mother's stomach stops producing digestive acid, allowing up to 30 tadpoles to grow. These live off their yolk sacs, and when they turn into froglets and the mother cannot stomach them anymore, she gives birth by simply opening her mouth.

The Surinam toad of the Amazon also has a bizarre method of parenting. This toad looks like it has been run over, with flattened legs and tiny eyes displaced to the front of its flattened, triangular head. But it is not just its appearance that is odd; its young get under its skin — literally.

The Surinam toad lives at the bottom of muddy rivers, surfacing only occasionally for a breath of fresh air. In the breeding season the water becomes a swirling mass of activity as males grab females around the waist and make their way to the surface. Each female releases around one hundred eggs, which the male fertilizes and spreads over the female's spongy back. Over the next day her back swells and starts to engulf the eggs scattered across her skin. Soon each egg is embedded in its own chamber, and the mother's back becomes a protective honeycomb. The eggs remain here for several months until the young have metamorphosed into tiny replicas of their parents. Then they begin to break through the membranous ceiling, but for about a day they remain half-emerged, with their arms waving in the air. When the froglets break free they leave their mother for good.

Most frogs leave their fertilized eggs to their own fate, and if a pond evaporates, many offspring perish. In Africa the searing heat makes this a common problem, but the male bullfrog is an attendant lifeguard. If the water level in their pool drops, bullfrog tadpoles call to their father for help. He responds to their click-like cries by finding the nearest alternative pool and digging a canal from it towards his troubled offspring. The sun-baked mud sets like concrete, but the bullfrog paddles away for hours to soften the surface, driven by the plaintive calls for help. When he finally breaks through, water rushes into the old pool and the tadpoles swim up the torrent to safety.

Like us, many higher animals have a womb to nurture their developing young. But it is surprising to find a similar feature in a fish, and it is even stranger

Above: Acting as lifeguard to his tadpoles, a male **bullfrog** digs a trench to supply water to his stranded young.

to discover that, in this case, it is the male that becomes pregnant. The fish that shows this role reversal is the sea horse.

A few days before mating, the male sea horse develops a "brood pouch" on his stomach. Then, after a series of courtship rituals involving color changes and synchronized swimming, male and female sea horses entwine, and the male's pouch meets the female's egg duct. She injects more than two hundred eggs directly into his pouch, where they are fertilized by his sperm. The pouch lining then becomes soft and spongy like a placenta, and begins to exude a nutritious liquid. For the next six weeks the developing embryos are nurtured, supplied with oxygen, and protected by the male; in effect, he is pregnant. When the baby sea horses are born the male even experiences contractions.

Among mammals it is extremely rare for the male to take a role in the birth of his offspring; only recently have humans become an exception to the rule. Now, in Western societies, men are often expected to attend the birth of their children, but in nature this practice is so unusual that only one other mammal does it.

Nature's ultimate New Man is the Siberian dwarf hamster. When his partner gives birth, the male Siberian dwarf hamster becomes an attendant midwife, showing all the care and concern expected of the role. Quints or sextuplets are usual, and, as each baby is born, the doting father licks them clean and cuts the umbilical cord with his teeth. Constantly watchful, he retrieves any newborns that wriggle too far from the nest. But within hours of the birth he reveals a less sensitive side to his nature, and sets about mating with the female again.

Below: A mother **opossum** suffers from overloading when her clinging family grows large.

Like us, many mammals look after their offspring until they can fend for themselves, but some have such large families that keeping track of the young can be a problem in itself. Baby opossums stay with their mother by hitching a ride and clinging to her fur with their teeth. This is difficult enough when the many youngsters are tiny, but when they have grown to nearly the size of their mother, she can hardly move.

The Asian shrew tackles the problem by keeping her young in line. When danger threatens, a youngster bites the fur at the base of its mother's tail and

Opposite: Young **Asian shrews** form a daisy chain behind their mother to avoid getting lost. If they lose contact they will reattach themselves to any moving object they see.

*...a whole daisy chain of **shrews** forms behind the mother. If she runs, they keep in step, playing follow-the-leader in a sinuous line.*

hangs on tight. Then another attaches itself to the first youngster's rear, and others take their turn in the line. Eventually a whole daisy chain of shrews forms behind the mother. If she runs, they keep in step, playing follow-the-leader in a sinuous line. They are so firmly attached that they will continue to hang on if the mother is picked up, dangling below her like a furry rope. The mother uses this linking technique to guide her young to safety, after which the line breaks up. But at the slightest sign of danger they instantly reattach and scurry away, like an attached group of schoolchildren crossing a road.

As well as reproducing, animals have to find food. The most extraordinary ways of doing this are found among predators, who use strength, speed, ingenuity, and a wealth of clever tricks to catch and kill their prey.

chapter three

Most predators come equipped with lethal weapons — for the **golden eagle**, a powerful hook bill is used to break into prey it has dispatched with its talons.

fantastic
feeding

Although we have the forward-facing eyes of a carnivore, and meat has always been an important part of our diet, we are ill-equipped for the job of killing our food. Even our canine teeth have been reduced to mere stumps, compared to those of our nearest relatives. Almost every other predator possesses fearsome weapons for killing prey, which makes us look strangely out of place in the natural world.

While we rely on ingenuity and tools, other predators have specialized, built-in equipment. But this specialization does not always work in the obvious way. In parts of southern Europe, golden eagles live mainly on tortoises, and no matter how well their talons are suited to catching and dispatching other prey, they are no match for an animal tucked resolutely inside its shell. The eagle's solution spawned a Greek legend. In the fifth century a soothsayer gave the poet and dramatist Aeschylus the dire news that he would die by a house falling on his head. On the appointed day, Aeschylus wisely ensured he was in open country in case an earthquake destroyed his house. It was in vain — a passing eagle fulfilled the soothsayer's prophecy by dropping a tortoise's house on the poet.

That the eagle's trick is more than legend has been shown by recent observations of golden eagles in Greece. In some areas tortoises make up more than 90 percent of the eagle's diet. To break into the shells, golden eagles pluck tortoises from the ground, carry them into the air, and drop them onto rocky ground. Egyptian vultures use the same strategy with ostrich eggs, and the lammergeyer breaks into the marrow of scavenged bones by dropping them onto rocks, too.

Above: The **tortoise**'s armour is a match for the **eagle**'s talons, so the eagle carries the tortoise aloft and breaks it open by dropping it on rocks.

Necessity has forced these birds to adapt their behavior in a way that reminds us of ourselves, and it is this that we find surprising. Most animals simply rely on the sophistication of their killing tools. Again, the strangest of these are those that remind us of our own inventions. One of the more extraordinary is a slashing saw.

The elongated snout…studded with vicious tooth-like projections…is used like a chainsaw to slice through shoals of fish.

Below: The **sawfish**'s elongated jaw is fashioned into a slashing saw that can lacerate its prey.

The elongated snout of the aptly named sawfish is studded with vicious tooth-like projections. The lethal weapon of this shark-like ray reaches 6.6 inches (2 m) in length, and is used like a chainsaw to slice through shoals of fish. The stunned or lacerated fish are then scooped up in its mouth, which is positioned conveniently below the saw.

Sawfish live in shallow tropical estuaries, where they also slash at the mud to flush out any hidden fish. This activity damages the teeth of the saw, but the teeth grow continuously from their base, and any broken serrations are soon replaced. The saw is not only used to find and kill prey — it also doubles as a formidable defensive weapon.

Another weird weapon belongs to the cone shells. These predatory sea snails inhabit tropical waters and detect their prey using an extensible tube that bristles with chemical sensors. The snail hunts so slowly that fish and other prey are rarely aware they have been targeted, and can often be approached to within touching distance. The cone shell then fires a harpoon into the prey's skin and injects a dose of powerful venom through the harpoon's hollow shaft. The lethal injection causes near-instant paralysis, giving the prey little chance to escape. The harpoon is joined to the snail via a thread, so the snail simply winds in its catch and engulfs it with its gaping mouth.

Mantis shrimps, or stomatopods, are also specialists in lethal weaponry. Depending on their type of armory, they either spear their food or smash it to death with heavily calcified clubs. The "smashers" pulverize the shells of hard-bodied prey such as crabs, lobsters, and snails. The "spearers" have lance-like appendages and specialize in soft-bodied prey, such as shrimps, fishes, and squid. Often the victim simply does not know what hit; the strike of a stomatopod is one of the fastest movements in the animal world. A spearer can accelerate its weapon from rest to a speed of 33 feet (10 m) per second in just

Opposite above: The **mantis shrimp** uses a club to smash its prey to bits, pulverizing its victim with a punch as fast as a speeding bullet.

Opposite below: The most complex targeting system of any animal: the central band of the **mantis shrimp**'s compound eye scans prey for color and polarization. The two eyes move independently and the bands act as cross-hairs; when they converge on prey, the killing clubs go into action.

4 milliseconds. It moves 10 times faster than the more famous praying mantis, despite having to punch through water, which is far denser than air.

One Californian species of "smasher" strikes with a force approaching that of a 0.22-calibre bullet, and can easily shatter double-layered safety glass. Not a good aquarium specimen, it can literally make a break for freedom.

Stomatopods couple their deadly weapon with a formidable targeting system — they possess the most sophisticated eyesight of any animal. While our eyes rely on just three types of photoreceptor cell to distinguish color, the mantis shrimp has sixteen. Four of these are devoted just to perceiving ultraviolet, a color we cannot even see. The combination of this targetting array with a weapon system that packs a killer punch probably makes the stomatopods, weight for weight, the most well-armed animals alive.

The **chameleon**'s tongue moves at ballistic speeds — the acceleration reaches 50 g — five times more than an F16 fighter jet.

Above: A **chameleon**'s extendible tongue secures its prey with the help of a tip that forms a suction cup.

Chameleons are more familiar high-speed assassins. New research into their extensible tongue has revealed just how ingenious it is. Many reptiles and amphibians use a projectile tongue to catch prey — among them is the Sardinian salamander, with a tongue that extends nearly the length of its body. The tongue's stickiness enables it to haul in small insects, but some of the larger chameleons will target bigger prey such as lizards, and even small birds. If they relied on stickiness alone to catch animals of this size, the tongue's tip would need to be as large as a baseball glove. But the chameleon has that problem licked. A couple of milliseconds before it hits, the tip curves inwards to form a suction cup. As soon as it makes contact, muscles in the tongue contract to draw the cup back, increasing the suction.

The chameleon's tongue moves at ballistic speeds — the acceleration reaches 50 g — five times more than an F16 fighter jet. The burst of speed is produced by spiral muscles in the tongue, which contract width-wise to make them stretch forward. A lubricant allows the muscles to slide at time-slicing speeds.

Similarly, squid and cuttlefish use rapidly elongating tentacles to surprise their prey. These accelerate with a force of 25 g. Although this is half the speed of the chameleon's tongue, the tentacles achieve their feat despite being handicapped by the resistance of the water.

Cuttlefish have other powers. They are renowned camouflage artists, able to vanish with a quick flush of their pigment cells. They also change color spectacularly when excited, flashing rapidly from yellow to reddish-orange and greenish-blue. These changes communicate mood changes, but when cuttlefish meet prey, their colorful displays serve a more fascinating function. Bands of color pass over the cuttlefish's body in a mesmerizing display that seems to confuse the victim momentarily; the cuttlefish's projectile tentacles do the rest.

Below right: The outer ring of the **cuttlefish**'s tentacles conceals a killing pair hidden inside. These can elongate with a force of 25 g to snatch crabs and other prey.

Above: **Stoats** sometimes appear to charm their prey by performing a strange, ritualized dance. They strike out at their unsuspecting audience with deadly results.

Hypnosis must rank as one of the strangest ways of getting a meal, but it seems that cuttlefish are not alone in using the technique.

The stoat charms its prey by performing a strange ritual outside a rabbit warren. It gyrates and springs into the air until it has gathered a fascinated audience of rabbits, who seem bewitched by the show rather than being frightened. As the stoat writhes and twirls, it sneaks closer to its unwitting prey. Then, without warning, it ends its show and dispatches a member of the audience with a bite to the neck. Whether the kill is premeditated or merely an accidental outcome of play is debatable, but most people who have seen the performance interpret the hypnotism as a deliberate and deadly ploy.

Another way of drawing prey into striking range is to use a lure. One of the strangest exponents of this method of hunting is the frogfish, or anglerfish. These weird, lumpy fish hardly ever swim; instead they crawl around the ocean floor using their pectoral fins as arms and hands. They come in almost every color but, despite their garishness, they are masters of disguise. As soon as they stop moving they vanish among the weeds, sponges, and corals. Rather than hunt down their prey, they prefer a more contemplative approach; like sport anglers they sit and wait with a fishing rod and bait. The rod is crafted from a modified dorsal spine that projects conveniently from the top of the fish's head. On the end is a fish-like lure that the frogfish wiggles enticingly over its mouth. The bait is a perfect replica of a fish, complete with convincing details like eyespots, fins, and a mottled pattern that mimics scales.

The mouth balloons to 12 times its volume, creating a huge suction pressure that pulls in the prey like a giant vacuum cleaner.

Once the lure has tempted a hungry fish into range, the frogfish makes its move. In a micro-instant, the frogfish expands its oral cavity and engulfs the prey. The mouth balloons to 12 times its volume, creating a huge suction pressure that pulls in the prey like a giant vacuum cleaner. Amazingly the whole process happens in just six thousandths of a second. This is so fast that a fish can be

Above: The
frogfish uses its
mouth as a vacuum
cleaner to suck in
prey faster than the
eye can see. A lure
on its head draws
the prey within
striking distance.

sucked out of a school without the other fish noticing, which gives the frogfish a sporting chance of snatching more victims from the same school.

The frogfish is not alone in using lures; the world's largest freshwater turtle —the alligator snapping turtle — has a pink, worm-like lure lurking in the bottom of its large mouth. The turtle lies camouflaged at the bottom of swamps with its mouth open and its worm-like bait wriggling. The interior of the mouth is grey and black, so the lure is all the more conspicuous. As soon as a fish tries to take the morsel the mouth snaps shut like a trap.

The South American horned toad shows a bizarre variation on the lure technique — it attracts prey by waggling its toes. But perhaps the most convincing lure of all is that of the Australian death adder, which is so perfectly camouflaged that it is almost impossible to see — except for the writhing tip of its tail. This wriggling appendage is yellow and segmented, just like the worm or beetle grub it imitates. It is irresistible to any insect-eating creature. Frogs, birds, lizards, mice, and rats all fall victim to the writhing, beckoning tail, and they are dispatched with a large dose of powerful poison. Unlike most snakes, the death adder rarely retreats from humans, preferring to rely on concealment. Consequently it is easily stepped on, with predictable and disastrous results. Although not a true adder, and more closely related to the cobras, it is well-named — its venom is a potent neurotoxin that rapidly cripples the nervous system. Half of all bites to humans are fatal.

Above: The **death adder** lures its prey by waving the tip of its tail like a writhing grub.

Not all lures mimic food — one of the most ingenious is a perfume. Many female moths release a type of pheromone to attract a mate from up to several miles away. The bolas spider synthesizes its own version of the scent and uses it as a fatal attraction. Camouflaged as detritus or bird droppings, it hangs on a twig and waits until a moth, attracted by the fake perfume, flies within range. Sensing the moth's beating wings, the spider unleashes its second deadly weapon — the bolas.

The bolas is named after a hunting weapon once favoured by indigenous tribes of South America. The device consisted of a rope with a number of weights at the end. It was whirled like a sling and then thrown at wild animals so as to wrap around their legs and trip them up. The only people who still use bolas today are the Gauchos, or cattle ranchers, of southern South America.

Below: The **bolas spider** swings a blob of sticky silk to ensnare male moths lured in by a perfume that smells like a female moth.

The spider's version of the bolas is made from a single thread of silk with a huge blob of sticky silk at the end. When the moth flies near, the bolas is swung enthusiastically until the moth is trapped. The spider then reels in the thread and eats its entangled prey.

Spider's silk is remarkable because of its unique combination of strength and stretch. It is squeezed from nozzles, known as spinnerets, as a liquid protein but immediately hardens on contact with air. The result is a thread as strong as Kevlar — the fibre used to make bulletproof vests — but far more elastic. Engineers have calculated that a woven cord of spider's silk as thick as a pencil could stop a jet in midair.

The circular "orb webs" are the best-known silk traps used by spiders, but there are many other designs. One of the less familiar is that woven by the net-casting spider. It holds a silk net in its forelegs until an unsuspecting insect passes by. Then, like a miniature gladiator, it hurls the net over its quarry. The spitting spider's trick is to spit a mass of sticky saliva that literally glues its victim to the ground. This glue-spitting technique reaches perfection among the velvet worms.

Velvet worms are thought to be closely related to the wormlike ancestors of the arthropods, the largest group of animals on Earth, which includes insects and spiders. Fossil evidence suggests that these multi-legged worms have changed little in millions of years. They can still be found today throughout the tropical forests of the southern hemisphere. Like spitting spiders they have the extraordinary ability to fire glue at their prey.

Velvet worms hunt other invertebrates among leaf litter on the forest floor. They shoot at prey with an adhesive liquid ejected by two nozzles next to the mouth. The nozzles move from side to side as they fire, causing the stream of glue to crisscross in a lasso-like motion. The glue travels nearly 3 feet (1 m) and dries in seconds, ensnaring the prey in multiple strands. With its victim pinned

Opposite: A **velvet worm** squirts streams of sticky "glue" to ensnare its prey.

down, the velvet worm bites a hole in the body and forces digestive saliva into it. The prey's insides liquefy, and the velvet worm sucks out the resulting soupy fluid. Velvet worms also use their glue guns as defensive weapons.

…another ace marksman…turns its mouth into a water pistol…. the **archer fish** *takes potshots at insects crawling on overhanging vegetation.*

The archer fish is another ace marksman, but its preferred ammunition is water. It turns its mouth into a water pistol by pressing its tongue against a groove in the roof of the mouth to form a tube. When the fish closes its gills, a jet of water shoots through the tube and up to 7 feet (2 m) away. In this way, the archer fish takes potshots at insects crawling on overhanging vegetation. It then skims its hapless victim from the water surface.

Previous pages: The **archer fish** uses its mouth like a water pistol to shoot down insects from overhanging vegetation.

The feat of shooting a stream of water to knock prey off a branch is remarkable enough, but the archer fish does it time and again with pinpoint accuracy. It accomplishes the trick despite the fact that light from the target undergoes refraction at the air–water boundary before it reaches the fish's eyes. Such bending of the light causes the prey to appear to be in a different location, but the archer fish makes appropriate corrections and even allows for the water jet to curve downwards slightly on its trajectory.

Although spitting water is the archer fish's favored method of catching prey, it sometimes leaps out of the water to pluck the insect directly from its perch. Leaping usually happens when the prey is close to the water, and if the leap fails, the archer fish reverts to using its water pistol. But some fish have taken the idea of leaping at prey to an extreme.

Arowanas are long, eel-like fish found in Australasia, South America, and parts of Africa. In Southeast Asia they are known as dragon fish because of their scaly skin and slender bodies. In England they are thought to ward off evil and are often kept as good-luck pets. Like mythical dragons they also take to the air, in this case to snatch insects from overhanging branches.

Opposite: The **arowana** makes spectacular leaps from the water to snatch insects from leaves and branches.

Right: **Crocodiles** can overcome their massive weight to propel themselves from the water at great speed.

Crocodiles are also expert leapers. Their feats are particularly impressive because of the handicap of their huge weight and the fact that their jumps begin from a stationary position. If the crocodile

is in the shallows, it uses its legs like spring-loaded pistons to push itself from the water at explosive speeds. The crocodile also makes high leaps from deep water to grab prey from overhanging branches; for this it undulates its tail with powerful contractions propelling it up to 8 feet (2.5 m) into the air.

Other predators can immobilize their prey without even making physical contact. One means of doing this, found only in the ocean, is to stun victims with high-intensity sound or pressure waves.

Above: **Killer whales** can stun fish by snapping their tail flukes close to shoals.

Working as a team, killer whales, or orcas, herd herring into a tight school close to the surface. Then they lunge at the fish with the underside of their tail flukes, creating a sharp retort that appears to stun the herring and makes them easy to scoop up.

…sound is concentrated in the forehead and fired like a shotgun at a shoal of fish.

Sperm whales and dolphins are believed by many to use sound as a weapon. The sound is concentrated in the forehead and fired like a shotgun at a shoal of fish; disorientated and stunned fish are then located by the whales' echolocation system. But the most stunning underwater killer has to be a type of shrimp.

The ocean is actually a very noisy place, and one of its noisiest inhabitants is the pistol shrimp, which makes sound by clicking its outsized claws together. The mass cacophony of thousands of these clicking pistol shrimps creates enough a din to drown out even the sound of a submarine. Remarkably, the noise is generated not by the claws themselves, but by tiny water vapor bubbles that form spontaneously and then collapse violently, triggered by pressure changes in the water between the claws. This process is known as cavitation, and the energy generated by it almost defies belief; the collapsing gas bubbles create explosive plasma balls that, for an instant, reach 5000°C — almost the temperature of the Sun.

The noise is used mainly for communication, but the high-intensity sound has another benefit. The sound waves are powerful enough to stun or even kill any nearby creature — just a quick snap, crackle, and zap and it's all over.

chapter four

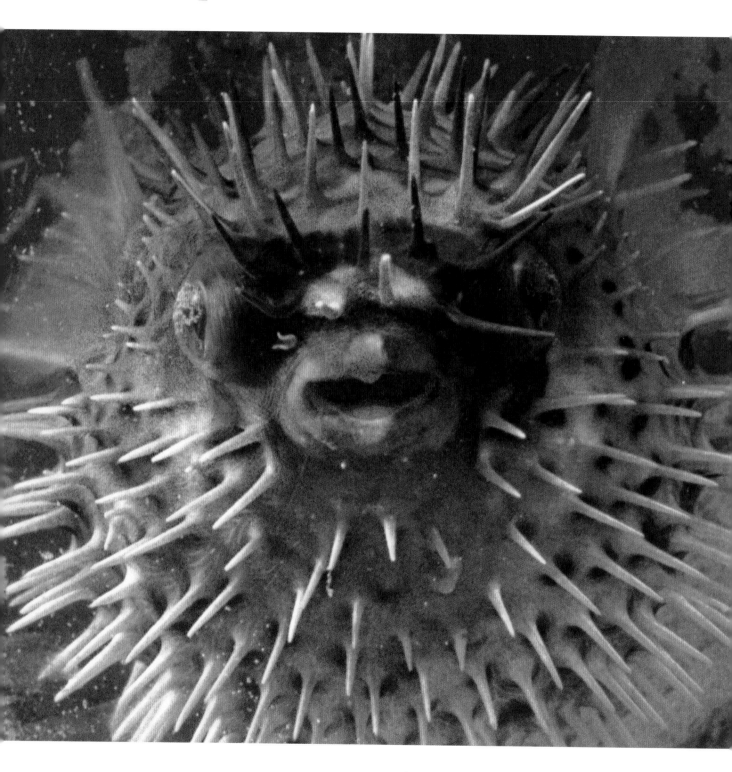

The poisonous **pufferfish** inflates itself with air or water to intimidate any opponent.

devious
defenses

In our natural state we suffer from a peculiar shortcoming — we are almost totally defenseless.

Fortunately, we have learned to compensate for this failing with an armory of defensive techniques

and devices. But our inventions are not as original as they might seem. Almost every form of self-

defense we use — including camouflage, armor, smoke screens, and even mace sprays — has a

natural counterpart. Many of these animal defenses make even our best efforts seem crude.

Camouflage is the most popular protection of all. The simplest version, employed by many

birds and mammals, and most fish, is countershading. The pale belly and dark back of a shark is a

typical example — it conceals the body by making the areas in shadow brighter so that they match

those that are in the light. Another popular ploy is to use disruptive patterns to break up a shape; the

spots of cheetahs and leopards work this way. All these techniques have military counterparts, but no

human design can beat nature's quick-change artists — the chameleon, octopus, and cuttlefish.

Fast color changes are practical only for animals that lack fur or feathers, as the changes involve the

skin. The technique relies on skin cells called chromatophores, which contain color pigments and

can quickly expand or contract. Octopuses, and cuttlefish flush with sudden waves of color as their

chromatophores dilate; stripes and strong patterns may be used to break up the body's outline, or more subtle shades may be used to blend into the background. These color shifts can take just two-thirds of a second.

Although chameleons are famed for their color-changing skills, they are influenced more by mood and temperature than by the color of their surroundings. The octopus reacts faster and blends in better.

Previous pages: The **leopard**'s spotted coat creates the impression of dappled shade and acts as camouflage. Body paint creates the same effect — there are two people hiding in picture.

A **peacock flounder** can mimic pebbles or sand

with such perfection that it appears to vanish.

Above left: This **leaf insect** uses its body to cover the hole it has eaten away.

Above right: The caterpillar of the **case-bearing clothes moth** weaves a wooly sweater to match the garment it has started to eat.

Opposite: A **peacock flounder** attempts to simulate the squares of a chessboard.

Of all the quick-change artists, flatfish achieve the most convincing match. They scrutinize their surroundings to work out the best disguise. A peacock flounder can mimic pebbles or sand with such perfection that it appears to vanish; put one on a chessboard and it will create a convincing facsimile within seconds.

To really disappear, an animal needs to be invisible; remarkably the glass catfish, an inhabitant of tropical rivers in Asia, almost achieves this spectral state. Like the invisible man, it lets light pass straight through its transparent skin and muscles. Strangely, when the fish dies its body becomes visible.

Another clever deception is to impersonate an inanimate object. Myriad stick and leaf insects play this cryptic game, some so well that they even include insect damage in their designs to avoid unconvincing perfection. Other impressionists include snakes that imitate vine tendrils, bugs that resemble

thorns, butterflies that copy dead leaves, and moths that look like patches of lichen. Even bird droppings have imitators — the Malaysian crab spider resembles a particularly runny specimen, complete with splash marks created from a thin film of silk.

If an animal's body cannot be shaped or colored to match the environment, the right hues and textures can still be acquired. Caddis fly larvae

Above: This **vine snake** imitates the curled tendrils of the foliage to conceal itself in the rainforest.

create mobile hideaways from the debris they find around them, while the

caterpillar of the case-bearing clothes moth knits its own matching wooly

sweater from the garment it destroys. But nature's ultimate seamstress is the

dressing crab.

The camouflaged suit of the dressing crab is fabricated from living

material taken off-the-rack from a coral reef. The crab uses its claws to carefully

snip off pieces of seaweed, sponges, or bryozoans, and apply them to its shell. The living fabrics are secured to the carapace by tiny hooks that work like Velcro, and the crab keeps fashioning its outfit until it has achieved a perfect match with its surroundings.

Another way to avoid being eaten is to take on the appearance of a more dangerous animal; every bee or wasp has a host of orange-and-black-striped impersonators. Snakes also play the impersonation game; the deadly Brazilian coral snake has a harmless doppelgänger with almost identical colored stripes.

A simpler ploy is to look like part of a dangerous animal. In one of the most wonderful impersonations in nature, the caterpillar of a Costa Rican moth has a false face that makes it look like a tiny viper. Many animals possess false

Above: The **dressing crab** decorates itself with fabric scavenged from its surroundings. If given pearls and lace it is equally happy to attach these unusual gifts to the Velcro-like hooks of its shell.

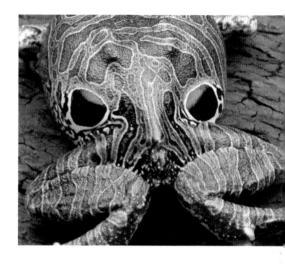

eyes, which either intimidate a predator so that it thinks it is facing a far larger creature, or fool it into attacking a less vital part of the body. The hairstreak butterfly sports a complete false head at the rear of its body, using tail filaments to double as fake antennae. When a bird attacks, the butterfly takes off in the opposite direction than expected, often leaving the attacker grasping at thin air.

The ultimate impostor has to be the newly discovered mimic octopus. Unbelievably, it morphs into other animals by manipulating its shape, color, texture, and tentacle positions. One minute it may flatten itself like a flounder and hide on the bottom, the next it may curl its arms and walk along the bottom like a crab, or it may even rearrange itself to look like a venomous lionfish. There have been reports of mimic octopuses shape-shifting into mantis shrimps, jellyfish, stingrays, and even sea horses. It may even become a medusa, with each tentacle resembling a banded seasnake — one of the most poisonous animals in the ocean. By mimicking a deadly animal the octopus achieves an effective defense but, bafflingly, it also copies harmless creatures. No one has yet discovered why.

One minute it may flatten itself like a flounder and hide on the bottom, the next it may curl its arms and walk along the bottom like a crab....

Even the common octopus has a gamut of defensive tricks — it can squeeze into impossibly small holes, it can produce a smokescreen of black ink, and, like many other animals, it can shape its body to look far bigger than it actually is.

Eagle owls are also inclined to puff themselves up when threatened — they arch their wings to simulate a huge and terrifying adversary. In a similar way, the Australian frilled lizard opens a cape-like ruff so that predators unexpectedly find themselves face to face with a creature of terrifying proportions. Likewise, when a toad meets a snake it inflates itself with air to appear too big to eat. In Central America, the horned frog goes one better; it adds an unearthly scream to its performance, which unnerves any predator not put off by the toad's over-inflated ego.

Above : The incredible **mimic octopus**, the ultimate impostor, here mimicking a lionfish and a blenny.
The mimic is on the right of each pair.

Opposite: An **eagle owl** and a **horned toad**, both puffed up and facing down the enemy.

Of all the puffed-up pretenders, the puffer fish take the crown. This odd, box-like creature crunches up coral with hard, beak-like teeth. To avoid becoming a meal itself, it quickly inflates with air and water, and transforms itself into a globe too big to swallow. At the same time, it erects poisonous spines to create the appearance of a particularly lethal pincushion.

Opposite: The
**poison-arrow
frog** warns of its
deadly secretions
with bright colors.

One of the puffer fish's poisons is tetrodotoxin, a venom so deadly that it is purported to be an ingredient in voodoo zombie potions. By careful preparation, the fish's poisonous parts can be removed. In Japan, where the puffer fish is known as *fugu*, it is a highly valued food. But it is a dangerous delicacy — if any poison remains, there is no known cure. Several diners die from its effects each year.

Being poisonous is the ultimate deterrent, but one that works only if an attacker reads the danger signs and does not kill. In nature a universal language of warning signals has evolved to prevent mistakes from happening. We use similar coded combinations of red, yellow, orange, and black in our danger signs.

The vividly marked, yellow and black poison-arrow frog is so deadly that South American Indians lace their arrows with the frog's skin secretions. The poisonous fire salamander displays the same warning marks on its belly, as do the bodies of many toxic beetles and moths. Tiger beetles opt for a red and black signal, and the familiar ladybird and the deadly coral snake use the same message.

It aims at the eyes in a jet that is effective

for nearly 8 feet (2.5 m); a direct hit causes

blindness and excruciating pain.

Instead of using a color code to warn of their potency, cobras spread out their hood. If this fails, the cobra follows up with a lethal bite administered through hypodermic fangs. The venom acts on the nervous system and causes

rapid paralysis usually followed by death. But close contact with an attacker is dangerous, so the spitting cobra dispatches its venom as a spray. It aims at the eyes in a jet that is effective for nearly 8 feet (2.5 m); a direct hit causes blindness and excruciating pain. The green lynx spider can also spit venom — a large female can hit an attacker 8 inches (20 cm) away.

Receiving a face-full of vomit must rank as one of the world's most unpleasant experiences, yet fulmars, whose name in Norse means "foul gull," specialize in this revolting defense. These sea-going birds leave their chicks for long periods while they forage over the ocean. Left alone, the chicks attract the attention of skuas, eagles, and gulls. But as soon as one of these predators approaches, the chick spews a volley of yellowish stomach oil into the attacker's face. Even hatchlings clambering from the egg can throw up on call; and at just

Above: A **spitting cobra** sprays twin jets of venom with deadly accuracy at the eyes of an attacker.

four days old they can hit an intruder 1 foot (30 cm) away. As they grow older, the chicks' vomiting power and accuracy improve, until nothing within 5 feet (1.5 m) is safe. If the first volley of vomit misses, the birds follow through with two or more additional blasts. Rock-climbers enter the fulmar's vomitorium with caution, knowing they will have to endure a barrage of projectile vomiting if they dare to scale the cliff in the breeding season.

The stomach oil has a repulsive sweet, fishy smell and causes feathers and hair to matt. The oil also destroys the feathers' insulation and waterproofing, leaving the attacker vulnerable to hypothermia or drowning. Even the chicks' parents initially receive the same repulsive welcome, but thankfully, after about three weeks recognition dawns, and the parents can approach less cautiously. An equally revolting defensive strategy is used by the fieldfare — but this time the other end of the bird is involved.

The fieldfare is a kind of thrush that nests communally in northern Europe. When a predator, such as a magpie, approaches a fieldfare colony, the birds scramble into the air and start an aerial bombardment of excrement. The projectiles are unleashed at the intruder with unerring accuracy. These splatter-gun attacks are no joke, as the feces destroy the water-repellent and insulating properties of the plumage and often kill the trespasser. For obvious reasons, each bird has a limited arsenal, so it pays them to nest together; up to 40 pairs may occupy one nest site. When one sets off on a bombing mission, a wave of reservists rapidly follows, creating a formidable no-fly zone. The strategy works — birds that breed communally rear more chicks than those that go it alone. In fact it is so successful that other birds, such as merlins, often nest among the fieldfares and rear more young as a consequence. Obligingly, they keep their side of the bargain by not eating their protectors' young.

Both the fieldfare and the fulmar recycle the body's waste products for use in self-defense. Other animals produce more specialized body secretions, but these can be just as obnoxious.

Skunks move around with a jaunty air of invincibility and announce their presence with a bold, black-and-white fur coat. They have good reason to be confident, for they are among the few mammals to possess a chemical weapon. Two glands just behind the tail manufacture a mace-like spray that can be directed at the face of any animal mugger. The spray is fired from nozzles in the skunk's anal glands, and the resulting jet can travel several yards. Even a predator approaching from the side is vulnerable, for the skunk can twist the nozzles in any direction. A direct hit results in temporary blindness and a stench so unbearable that it frequently induces vomiting.

Skunks try to avoid resorting to their chemical weapon. The black-and-white coat acts as a warning, and if that fails, a skunk will wave its bushy tail like

Opposite: The **spotted skunk** performs acrobatic handstands as a warning. If this fails to deter the intruder, it sprays an excruciating stench.

When threatened, it stamps its front feet,

backs away, and throws its rear end over its

head. Then, like a break-dancer, it walks

around in a handstand.

a flag. The spotted skunk has even more bravado. When threatened, it stamps its front feet, backs away, and throws its rear end over its head. Then, like a break-dancer, it walks around in a handstand. It only unleashes its chemical deterrent if this acrobatic performance fails to impress.

Perhaps the most disturbing sprayer of all is the horned lizard of North America's southern deserts. This lizard actually squirts blood from its eyes. To perform this horrific feat, the lizard increases the blood pressure in its sinuses until their walls burst. Blood then jets from the eyes with such force that it can hit a predator 3 feet (1 m) away. Although the blood is harmless, the assailant is usually so stunned by the gruesome display that it leaves the lizard alone.

The spray of the bombardier beetle is almost as shocking, for it features a controlled chemical explosion. The armament is housed inside the beetle's rear where there are two chambers of chemicals: One contains a mixture of hydraquinine and hydrogen peroxide, and the other contains an activating agent. The chemicals are kept apart until danger threatens; then the beetle opens a release valve and discharges the ingredients into a combustion chamber. The result is explosive. A boiling spray of noxious chemicals is propelled through a nozzle at the tip of the beetle's body. The nozzle is directional, allowing the corrosive bombardment to be aimed at attacks mounted from any direction.

The bombardier beetle's multichambered explosive system resembles the rocket technology used in Germany's V1 missiles, or "Flying bombs," during World War II. In both systems the explosion is controlled by releasing the chemical ingredients in pulses, rather than in one explosive burst. The beetle does this 735 times each second; the flying bomb could manage a pulse rate of only 42.

Another chemical defense is used by the hagfish. This almost blind, eel-like fish lives in the cold oceans. It feeds by rasping the flesh from dead or dying fish that fall to the muddy bottom. Its strange, jawless mouth is surrounded by sensory tentacles and contains a tongue-like projection studded with teeth. Even stranger is the hagfish's way of defending itself. If attacked, it slimes its assailant to death.

When the hagfish is provoked, slime oozes from the hundred or so glands that line its body. The secretion swells in contact with sea water until it forms a slimy cocoon around the fish. The glands also secrete fine fibers that reinforce the slime and allow it to stretch without breaking. Each thread is more than 2 inches (50 cm) long, but the hagfish employs a neat method to prevent tangling — the threads are played out from the glands in winding layers, like a nautical rope stacked into a figure eight.

The reinforced slime protects the hagfish from attack and may even suffocate predators such as moray eels. The hagfish avoids becoming trapped in its own slime by a clever trick: It simply ties itself into a knot and wipes the slime away by passing the knot down its body. An adult hagfish can produce a prodigious quantity of slime — enough to transform a bucket of water into a mass of viscous jelly.

Below: When provoked, the **hagfish** secretes slime to form a protective cocoon around itself.

*The **three-banded armadillo** possesses the ultimate design; its armored head and tail mesh together to form a perfect and invincible ball.*

For their defense, tortoises and turtles prefer full body armor but, like medieval knights, they suffer from problems of maneuverability. The armadillo's solution is to use articulated armor plating. Its protective shields consist of a double layer of horn and bone that covers most of the exposed parts of the animal. Hinged bands allow the armor to bend so that the armadillo can curl up to protect its vulnerable underbelly. The three-banded armadillo possesses the ultimate design; its armored head and tail mesh together to form a perfect and invincible ball. The animal can check for danger through a small chink in its armor but, when touched on the chest or belly, it snaps the edges together like a steel trap.

In Africa the pangolin takes a similar approach to armor plating, but it opts for a design of many overlapping shields that give the appearance of huge scales. These allow more flexibility.

As everyone knows, hedgehogs and porcupines defend themselves with an armory of spines, but less familiar is the sharp-ribbed salamander of Japan. This amphibian uses body piercing as a defense — it contorts its body when attacked so that the ribcage flattens and the tips of the ribs shoot out through small apertures in the skin, making it a spiky mouthful for any predator. The ribs are also laced with a mild poison.

A further variation of self-mutilation is practised by many lizards — they willingly give up their tails to predators. Lizards can perform this grisly

Opposite: The **three-banded armadillo** can roll itself into a perfect, inpenetrable ball.

amputation because their tail vertebrae have built-in fracture points. Like delicate glassware, the tail of the legless lizard, known as the pallas glass snake, fragments into many pieces when handled roughly. The muscles continue to contract, however, giving the severed pieces a life of their own, which unnerves any predator. In the Australian chameleon gecko the tail squeaks as it writhes around, causing even greater consternation.

Another way of distracting a predator's attention is to feign injury, a common strategy employed by birds. This behavior always happens near a nest, with the aim of drawing the predator away from eggs or young. Pipits, for example, flutter in front of would-be egg thieves, looking like an easy target. But as soon as the predator gets close, the pipit stages an instant recovery and flits away. By repeating the masquerade, the parent soon leads the danger away. Many plovers practise a more elaborate technique; they let a wing droop as if it were broken, and lead the predator away from their young.

The greatest bluff of all is to feign death. This exploits the fact that many predators will refuse carrion, or only recognize prey if it moves. Caterpillars, beetles, spiders, and many other animals are all exponents of this macabre trick. Toads lie totally motionless on their back if attacked by an animal or if held by a person for any length of time, and birds such as pigeons can be put into a death-mimicking trance simply by tucking their head under a wing. Likewise, chickens can be immobilized by placing them on their back and drawing a line in the dirt away from their head, and alligators simply need to be turned upside down and held briefly.

Acting dead is common throughout the animal kingdom, but there are some animals that give a more polished performance than others. Among these are the hognose snake and the opossum.

The American hognose snake goes belly-up when attacked and lets its body turn limp. For extra realism it lets its tongue loll out of a gaping mouth, and

Above: These **doves** are in a death-mimicking trance, prompted simply by tucking their heads under a wing.

Opposite above: The American **hognose** feigns death by going belly-up and letting its body go limp.

Opposite below: Literally "playing possum." An **opossum** feigns death by rolling over and lying still with mouth and eyes half open.

to gild the lily it releases a foul-smelling fluid from its anus. Unfortunately, if turned over, it instantly spoils the performance by rolling back to confirm it really is dead.

The opossums of North America are such experts in the art of feigning death that they have given rise to a popular expression. "Playing possum" actually involves quite the show — when danger threatens, the opossum first tries bravado, snarling with teeth bared and hissing loudly. If this bluff fails, it falls on its side and lies still. With gaping mouth and eyes half-open in an expressionless stare, the opossum appears dead to the world. For a finishing touch, it defecates and releases an evil-smelling green slime from its anal glands, in mimicry of a rotting corpse.

The opossum's act may continue for up to an hour. Periodically, the opossum checks to see how the performance is going down with the audience. As soon as the predator becomes bored and moves away, the opossum rises from the dead and scurries off the stage.

chapter five

Playing **cat** and **mouse**. The cat became a partner with people in the granaries of the Middle East, where it acted as a mouser. If the two are reared together they can become friends.

puzzling
partners

Another of our strange traits is the number of partnerships we have struck up with other animals. The dog, which is now such a part of our lives, was domesticated from the wolf between 12,000 and 15,000 years ago as an aid to a hunting way of life. Goats and sheep were domesticated around 10,000 years ago when people first settled to grow crops, and cattle were domesticated later. Cats were tolerated as mousers in the early granaries before they took on an additional role as companion animals. Pigs, donkeys, horses, and camels were soon added to the list and, over time, a menagerie of other animals, including llamas, water buffalo, chickens, and reindeer, were brought into the human fold. Even the guinea pig, an animal that we dismiss today as a child's pet, was domesticated for food more than 6000 years ago in Peru.

In some cases the association grew naturally into a stronger relationship, but often there was a deliberate attempt to control and breed the animals concerned. The end result is that both ourselves and our partners have become some of the most numerous animals on the planet. Some species have even avoided extinction; the auroch — the wild ancestor of cattle — is long gone and the wild ancestor of the horse is also extinct. Despite the apparently one-sided nature of our relationship with domestic animals, biologically speaking, the animals have forged a successful partnership, since their populations have increased along with their range. Similarly successful relationships between species are found throughout the natural world.

In the gardens of almost every home, herds of miniature domesticated animals are reared and milked for food. Usually we fail to notice this pastoral scene, as the farmers concerned are ants and their flocks are greenfly. As these aphids suck up sap from our prize roses, they store a sugary treat for the ants. The ants milk the aphids for this "honeydew" by stroking them with their antennae. The ants even defend their herds against ladybirds and lacewings, like herdsmen protecting their livestock from wolves. If the aphids are in danger of overgrazing, the ants move them to new pastures. The ants that tend the bean aphid even overwinter their herds in underground stalls, a form of transhumance like that practised by cattlemen in the Swiss Alps.

Below: The earliest farmers. **Ants** "milk" and tend to their herd of **aphids**.

The Australian green tree ant has struck up a similar relationship with the caterpillar of the common oak blue butterfly. The caterpillar has a nipple on its back that works like a cow's udder. When ants milk it, the nipple secretes honey dew. At night the ants use leaves to build a stall for the caterpillar and, during the day, they act as herdsers as it grazes. If marauding spiders or wasps approach, they are chased away by shots of formic acid fired by the ants. The ants' husbandry does not encompass other caterpillars, which are mercilessly consumed. To avoid this fate the oak blue identifies itself with a low rumbling vibration and a characteristic scent.

Above: This **screech owl** might catch a **blind snake** for its prey. If the snake escapes into the owl's nest, it will become an unplanned pest controller.

The partnership between ants and aphids or caterpillars is mutually beneficial and complex. Like many partnerships, it probably began by accident. There are present-day examples that show how such relationships may begin.

In Texas screech owls prey on snakes and bring them back to the nest to feed to their hungry chicks. Sometimes the owls catch a small burrowing snake known as a blind snake. Because it has slippery skin, the blind snake often escapes before disappearing down a chick's gullet. As it burrows into the debris of the nest, the snake finds a ready supply of maggots and nest parasites. Gaining free board and lodging, the blind snake inadvertently becomes a pest controller and helps keep down the number of parasites. The relationship is almost certainly accidental but both parties benefit.

*Although **tarantulas** occasionally eat small frogs they seem to know that their roommates are off-limits, and often a **frog** can be found nestling beneath the spider's hairy legs.*

A frog and a tarantula have struck up a similar partnership in the deserts of Texas. The tarantula digs a burrow for shelter during the day but soon some frog-eyed boarders move in as well. Although tarantulas occasionally eat small frogs they seem to know that their roommates are off-limits, and often a frog can be found nestling beneath the spider's hairy legs. For its rent, the frog consumes any ants or other insects that try to eat the spider's eggs, catching them with a flick of its tongue. In this way it gains free board as well as lodging.

The tarantula also provides the frog with security. As well as a venomous bite, the spider has a novel means of defense. By rubbing its feet across its furry back it unleashes an arsenal of arrow-like hairs. These are fired at the eyes of any intruder unwise enough to enter the burrow. Barbs on the hairs make them very difficult to remove, and even if they miss their target, the hairs readily fragment and can be inhaled instead. Either way, they afford the frog all the protection it needs.

The spiny-tailed lizard of the Arabian Desert also employs a personal security guard. Like the tarantula, the lizard digs a burrow to protect it from the desert sun, but it can still be attacked by animal predators or human hunters who value it as a source of food. Fortunately, scorpions also regard the burrow

Above: The **frog** deters ants from the **tarantula**'s egg case while the spider provides a home and security service.

as a desirable residence, and the two have struck up a bargain. The lizard tolerates the presence of the scorpion, which never stings its housemate, and in exchange for a home the scorpion acts as security guard — any predator that enters the burrow soon discovers that the lizard's housemate has a sting in its tail.

Hermit crabs are soft-bodied crabs that live inside discarded seashells. For additional protection, some hermit crabs appoint a sea anemone to be their personal bodyguard. The crab delicately uses its pincers to remove the anemone from a rock and plant it on its shell. The anemone's tentacles are armed with poisonous stings, so fish and octopuses that might normally eat the crab are put off by the possibility of being stung. The crab so values the anemone's protection that it takes the anemone with it when it outgrows its shell and has to move to a new home.

Predators reckless enough to approach receive a fistful of stinging tentacles.

In the Indian Ocean there is a crab that goes one better and carries anemones around in its claws to use as a defensive weapon. Predators reckless enough to approach receive a fistful of stinging tentacles. The crab's poisonous fisticuffs are so vital that its pincers have evolved a special shape to carry anemones. As a consequence, the next pair of legs are converted into tools for mincing up food.

Clownfish also profit from the anemone's protection racket. These gaily painted fish take refuge among the poisonous tentacles of large sea anemones and are immune to the stings. They are poor swimmers and would not survive for long away from the anemone's stinging defenses. The anemone also offers a secure nursery for clownfish eggs. In return, the clownfish act as valets, cleaning the anemone and eating parasites. As they dart between the tentacles, the clownfish also improve water circulation. The anemones also gain by scrounging leftovers from the fishes' meals.

Above: A **Boxer crab** with its pincers full of anemone tentacles.

In Australia the protected turns protector: Butterfly fish, like clownfish, are also immune to anemone stings, but they feed on the tentacles. However, if a clownfish is in residence it confronts an intruding butterfly fish with bared teeth — an act of bravado that usually drives the attacker away.

Surprisingly for such a close relationship, the clownfish is initially vulnerable to the anemone's poison. The first time a clownfish enters an anemone it approaches carefully to avoid too many stings. It returns repeatedly to its host, swimming in an elaborate dance among the tentacles and letting them touch its fins and body. Over time, it slowly becomes immune to the poison. The fish gains its immunity from a coating of mucus that covers its body. During its dance, the clownfish spreads mucus onto the anemone's tentacles, and, in turn, mucus from the tentacles coats the fish. The clownfish's slime appears to store a chemical used by the anemone to prevent it stinging itself when two tentacles touch. As a result, the fish achieves complete protection and the anemone gains a cleaner.

*The **cleaners** will boldly enter the gaping mouth of even a huge **grouper** but the grateful customer never snaps its mouth shut at the wrong moment.*

Personal hygiene is so important in the ocean that coral reefs have areas set aside as beauty parlors for fish. Here a barbershop of attendants, known as cleaner fish and cleaner shrimps, line up to tend visiting fish. The cleaners are recognized by their brightly striped uniform; sometimes the stripes are blue and white, but, appropriately, the cleaners often sport the familiar red-and-white trademark of a barber's pole.

The barbers bob and fuss in front of their customers to signal their willingness to attend. The visitors ask for service by suspending themselves in the water, either head down or head up, and opening their gills and mouth. The cleaner shrimps and cleaner fish then go to work. They tidy up their client by trimming off any loose skin and snipping away at fungal growths with their mouths. They also remove any fish lice infesting the skin.

The cleaners will boldly enter the gaping mouth of even a huge grouper but the grateful customer never snaps its mouth shut at the wrong moment. Not all the attendants are to be trusted, however. Some fish wear the uniform of a cleaner but have a very different code of practice. Instead of gently removing skin flakes and other debris, the sabre-toothed blenny dives at its customers and slices out chunks of flesh with its teeth, like a cutthroat barber wielding a razor blade.

The benefits of a personal valet are also appreciated on land. In Africa oxpeckers scurry over buffalos, giraffes, rhinos, and other big game to pick off fleas and ticks. Their roster of services includes special attention to the ears and eyes and an intimate probing beneath the tail. The oxpecker is well-suited for the task. Its

Above: **Cleaner shrimp** adopt the red and white markings of the barber shop pole and provide a cleaning service to fish like this **orange-spotted cod**.

beak is flattened to probe for ticks beneath fur; its toes are elongated to provide a secure foothold; and its tail is stiffened to form a prop that helps the bird clamber over an animal's flank. The only downside to the relationship occurs if the host has a wound; the oxpecker is quite happy to complement its meal with a side dish of blood. This vampiric tendency is particularly evident when the host is a domestic cow, which suggests that the oxpecker's partnership with this relatively new animal has not had time to become as well-balanced as that with wild animals.

Finches in the Galápagos Islands provide a similar valeting service for giant tortoises. The reptile solicits attention by craning its neck and stiffening its legs to lift the shell clean off the ground. In this way, all the nooks and crannies of the tortoise's wrinkled skin are exposed to allow a thorough grooming.

Keeping body and home clean is so important that many Central American mice and rats travel around with their own team of pest-control agents. Forty types of rove beetle provide this service for a variety of clients. The beetles can be nearly a 10th the length of their host, yet a mouse may carry as many as 10 of these bruisers clustered around the neck like a necklace. Whenever the mouse is on the move, the beetles' large jaws clamp firmly to the base of the fur. During the day, when the mouse returns to its burrow, the beetles disembark to do their work: their job is to hunt down the fleas that plague the mouse's nest. This pest-control service must be very important, since the healthiest mice are usually those with the greatest number of hangers-on.

The three-toed sloth of South America carries a less useful living cargo. This languorous creature crawls so slowly through the tree tops that it is often colored green by algae growing on the surface of its shaggy coat. The algae help camouflage the fur, but sharing the ride is a busload of small moths that provide the sloth with no particular benefit. There may be as many as one hundred of these hitchhikers on board, each waiting for the rare moment that the sloth defecates. Amazingly for such an inactive animal, the sloth travels all the way to the ground to use a special midden, or dunghill, for the purpose. The

below: The slow moving **three-toed sloth** carries around a busload of hitchhikers.

Opposite: The **moths** that use the sloth's body as a taxi service.

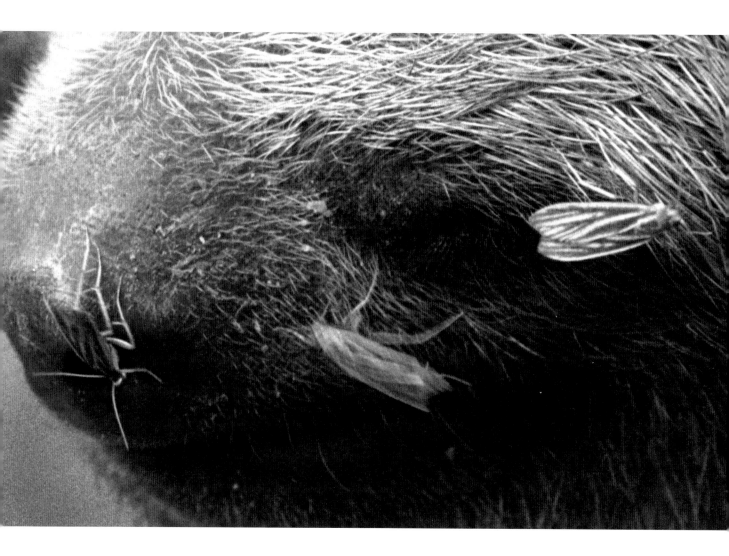

moths then jump off and lay their eggs in the steaming dung, which provides food for the caterpillars and, later, a place for them to pupate. Before the sloth leaves, any newly emerged moths — the outcome of previous visits — hitch a ride on its fur. They mate on this moving shag rug and disembark about a week later at the next midden.

In the same region, flower mites use hummingbirds as a private plane service. The flower mites feed on pollen and nectar in newly opened flowers. To reach new flowers or different plants they need to make a flight. As soon as a hummingbird visits their flower, the flower mites race up the bird's long bill as fast as their tiny legs can carry them. In the rush to be first onboard they can move at 12 body lengths per second — equivalent to a cheetah running at full speed. The reason for hurrying is that the mites have to reach a safe compartment — the hummingbird's nostrils — before the bird takes off. On an average flight, up to 15 stowaways may be safely tucked onboard. The mites are not particular about which airline they fly — any species of hummingbird will do — but each type of mite chooses to disembark at its own specific flower. The bird gains nothing from these travellers, but they do it little harm. Other animal passengers wreak havoc while onboard their host, however. These are the parasites.

Parasites have such weird and complex lifestyles and come in so many designs that they really require a book of their own, but one deserves special mention because the way it affected its host caused a baffling phenomenon.

During the 1990s, in the southern states of the U.S., frogs started to appear with extra limbs and other deformities. The most severely affected was the pacific tree frog – some were found with as many as 10 legs. Radioactivity and industrial pollution were blamed for triggering these genetic deformities until the real culprit was discovered. It turned out to be a tiny flatworm, known as a trematode, that spends the first stage of its complex existence in the body of a heron.

When a heron drops feces containing trematode eggs into water, the eggs are often eaten by snails. Once inside a snail, the trematodes hatch and grow. Eventually they burst out of the snail and swim towards the nearest tadpole. With torpedo-like precision, they aim for the limb buds and burrow inside. Here, the effect on the developing limb cells is like an explosion, and as the damage is repaired errors inevitably occur. Often each fragment of limb bud starts to grow anew, creating several versions of the same leg. When

...the **mites** have to reach a safe compartment — the **hummingbird**'s nostrils — before the bird takes off. On an average flight, up to 15 stowaways may be safely tucked onboard.

Below: **Flower mites** boarding their own private flight — in the rush to reach the shelter of the **hummingbird**'s nostril, they reach the equivalent speed of a running cheetah.

these multi-legged frogs mature, their deformities make them particularly vulnerable to attack by herons, increasing the chance that the parasite will end up back in its primary host.

     While some trematodes manipulate their hosts' anatomies, others can take control of behavior. *Dicrocoelium* is a trematode that spends part of its life cycle in a bird and the rest in a snail. It changes the snail's behavior so that it stays out in the open after dark, making it more likely to encounter a foraging bird. To ensure the snail gets seen, the parasite invades an eyestalk. The mass of swarming trematodes make the tentacles pulsate like green neon signs, sending out a message that has only one meaning — "Eat me."

     The worst thing that can happen to a parasite is to end up in the wrong host, and one of the worst things that can happen to a human is to become the mistaken host of the Candiru fish. This parasitic fish is found in the rivers of the

Below: Seen from underwater, this **candiru fish** finds its host by following a trail of uric acid, a technique that can be devastating when it hones in on the wrong target.

Brazilian rainforest. It usually inhabits the gills of other fish and it finds a new home by tracking a trail of uric acid. Unfortunately, a person relieving himself in the water gives off the same signals, and the fish quickly swims up the urine trail and into the urethra. Then it makes itself noticed by opening up backward-facing spines. The only way the fish can be removed is by surgery.

Most parasites that make their home in our bodies are much better-suited to the role of tenant — it is even possible to harbor a tapeworm 16 feet (5 m) long without noticing side effects. Although not all of us have such a spectacular companion, whether we like

Above: **Leeches** have made a medical comeback — they can be used to reduce swelling.

it or not we become a living city to other organisms. Countless bacteria swarm over our skin, and millions of microorganisms inhabit our intestines; follicle mites dwell at the bottom of our eyelashes, and lice are still a common problem among schoolchildren. Fortunately, most of these tenants do no serious harm, and some are essential for our survival.

Even leeches, once valued as a medieval cure for a whole range of ills, have recently made a medical comeback. Their ability to reduce swellings by sucking blood through the skin makes them useful in the treatment of cauliflower ear and other types of haematoma (blood-filled swelling). When the multitude of human parasites is added to the list of animals we have domesticated, the total

number of animal partners that it makes
is probably greater than for any other
creature. And there are still more to
consider.

In Africa many local tribes have
established an astonishing relationship
with a bird known as the greater
honeyguide. In an incredible feat of
memory, this bird makes reconnaissance
flights over an area as big as 100 square
miles (260 square km) and memorizes the
location of every bee's nest within it. Then
it looks for a person to open one of them.

As soon as the honeyguide hears
human voices, it flies towards them.
Perching in a nearby tree, it uses a
persistent, double-pitched call to attract
attention. Then, flitting from branch to
branch, it waits for a reaction. A human
honey-gatherer knows the correct
response is to move towards the bird.
The bird replies by flying away, using a

conspicuous, undulating flight. In dense scrub it quickly disappears, but it is
rarely gone for long — it is simply checking the position of the nearest bee's nest
before flying back to the gatherers. If they are still following, the bird flies off
again in the direction of the nest. This to-ing and fro-ing may continue for several
miles until the men are drawn close to the nest. The honeyguide then perches

Above: The **honeyguide** leads the honey-gatherers to their goal and waits for its reward.

nearby and emits a softer, less persistent call to signal the end of the journey. The bird waits patiently until the men have broken into the nest and gathered the honey. The men leave a thank-you gift of grub-filled honeycomb as a reward.

Such a perfect partnership between people and wild animals is unusual, but not unique. In some parts of the world a cooperative fishing venture has developed,

and it is every bit as remarkable. The custom occurs in parts of the Mediterranean and Mauritania, but it reaches perfection on the beaches of Laguna, near the southern tip of Brazil. Here fishermen stand in line on the shore ready to throw their nets on a cue from a dolphin.

In the murky water the fish are hidden from human view, but they are easily detected by the dolphins' sonar. When the dolphins find a shoal they signal its position to the fishermen. Then the dolphins herd the fish towards the shore,

Above: In Laguna, Brazil, **dolphins** like these herd fish towards the nets of waiting fishermen.

using a lunging dive to tell the fishermen exactly when to cast their nets. The dolphins' reward is the chance to catch mullet panicked by the encircling nets.

This cooperative fishing began several hundred years ago, and young dolphins learn the technique by watching their parents. Likewise, the fishermen pass on their knowledge to their children, and every dolphin is given a name. In this way the techniques involved are handed down, in the families of both humans and dolphins, from one generation to the next.

chapter six

Pregnant **elephants** will walk many miles out of their way to find leaves of the *Boraginaceae* tree, which help to induce overdue babies.

peculiar potions

Modern medical knowledge is so highly developed that it would be natural to assume we have little to learn from the natural world. But recently there has been renewed interest in herbal remedies and, even in conventional medicine, around 60 percent of all drugs are wholly or partly derived from natural cures. Aspirin comes from willow bark, which, in its unprocessed form, was a traditional treatment for headaches. Digitalis is derived from foxgloves and is still used to treat congestive heart failure. Quinine, a malaria cure from the bark of the Cinchona tree, has staged a comeback as mosquitos develop immunity to more modern drugs. Yew, a tree that has been revered for thousands of years, has become a source of the drug tamoxifen — a treatment for breast cancer.

As modern medicine rediscovers many ancient remedies and searches traditional cultures for other beneficial plants, a new branch of knowledge, known as zoophamacognosy, is also exciting interest. It reveals that we are not alone in understanding the medicinal power of certain plants.

In the forests of Tanzania, chimpanzees suffering from diarrhea treat themselves with a shoot from the Mujunso ("bitter leaf") tree — they strip the leaves and bark to extract the healing, bitter pith.

When ill, people in the nearby Watongwe tribe use the same plant to brew up a curative tea. In both cases, a cure usually takes less than 24 hours. The active ingredient in the plant has been shown to be effective for over 25 common ailments that affect people, half of them intestinal and parasitic.

Chimpanzees use the leaves of the *Aspilia* plant, a relative of the sunflower, for similar medicinal purposes. The leaves, which are covered with short, bristly hairs, are swallowed whole. The chimps will carefully fold the leaves before they eat the hairy mouthful. It seems the hairs somehow hook or trap the worms as they pass through the gut.

In South America, tamarins, infected by parasites, use a similar purging treatment, gulping down huge seeds that are almost too big to swallow. These medicinal gob-stoppers destroy anything in their path as they make their way through the monkey's digestive tract.

Chimps and gorillas know 30 different hairy-leafed plants that can physically cleanse their systems. Local people make preparations of the same plants to cure ailments such as dysentery and malaria, so it seems the leaves probably give a double whammy to the ape's parasites by acting chemically, too. Tribal cultures are aware of many other animal medicines besides this herbal diarrhea cure.

Kodiak bears have been seen chewing the root of *Ligusticum* (a plant of the carrot family) and then spitting the resulting mixture of saliva and juice onto their paws and applying it to their fur. The Navajo Indians consider the plant to be their chief medicine and according to their folklore they learned of the plant's healing qualities from the bear.

When pregnant Kenyan woman are overdue, they sometimes brew up a tea from the leaves of the *Boraginaceae* tree. Significantly, a pregnant elephant has been observed going out of her normal way and walking 17 miles (27 km) to consume a tree of the same species, just prior to giving birth.

Above: A **tamarin** looks for large seeds to clear its digestive tract of parasites.

East African tribes use the leaves of *Combretum* and *Ziziphus* as a primitive means of birth control to induce abortions. Chimps have been observed taking the same leaves, possibly as a method of controlling their own numbers.

It appears that birth control may be common among primates. At certain times female muriqui monkeys in Brazil eat a fruit from a plant known as monkey's ear, which contains stigmasterol, a chemical similar to the pregnancy hormone progesterone. At other times the monkeys eat leaves from plants containing compounds similar to oestrogen, which decrease fertility. Baboons even have a cure for period pains; the leaves of the African candelabra tree appear to ease their discomfort.

As well as coping with illness and fertility problems, animals need to deal with the poisons that lace much of their diet. These tannins, alkaloids, and other dangerous chemicals are plants' natural defenses against being eaten. Indian almond and mango trees use phenol compounds, but in Zanzibar their toxic leaves are the prime food of red colobus monkeys. To avoid severe indigestion the monkeys take charcoal as a preventive medicine . Doctors prescribe the same remedy for poisoning by the deathcap mushroom. The monkeys treat the stores of forest charcoal-burners as their pharmacy and carry out raids for their daily dose of medication.

Tropical butterflies suck up mineral-rich river water…while simultaneously squirting the processed desalinated water from their rears.

Another doctor's treatment for stomachache is a type of refined clay known as kaolin. Chimpanzees, gorillas, and various monkeys have all been seen dosing themselves with natural versions of the cure. In South America, spider monkeys eat dirt so enthusiastically that they resemble mud wrestlers as they scramble around in the local clay pit. Nearby, a kaleidoscope of colorful macaws and other parrots sort out their own gastric problems by nibbling away at the exposed clay in river banks. They are joined by peccaries, deer, and a variety of tropical birds. Even the forest people deliberately ingest clay when eating wild roots and tubers to neutralize the poisons such foods contain. As well as using the clay as medicine, animals gain valuable dietary minerals from it.

Above:
Butterflies take their mineral supplements — they suck up water from the riverbank to extract vital salts.

Most animals know when they need a mineral tonic, and seek out salts, such as potassium and sodium, to correct any body imbalance. Tropical butterflies suck up mineral-rich river water to satisfy their requirements, while simultaneously squirting the processed desalinated water from their rears. Natural mineral seams soon become sites of pilgrimage for many animals. Herbivores, in particular, are drawn to these salt licks. Elephants crave salt so much that, in one locality in Kenya, the work of countless tusks, drilling into the face of a cliff over the centuries, has created huge mines that penetrate the cliff for 660 feet (200 m).

In addition to taking vital minerals, animals self-prescribe vitamin supplements. When a rabbit licks its ears it not only cleans them; oil secreted onto the ears contains a chemical that breaks down in sunlight to form vitamin D. The rabbit gets a vitamin dose every time it cleans its ears.

…some birds appear to indulge in aromatherapy … nestlings that hatch into herbal nests cope better with stress and put on weight faster….

As well as taking medicines, mineral supplements, and vitamins, some birds appear to indulge in aromatherapy. Male starlings weave sprigs of fresh herbs into the dry leaves and twigs of their nest. Many of these, such as lavender, thyme, and peppermint, provide the essential oils used in aromatherapy. Recent research suggests the connection is far from accidental. The herbs benefit the starling's immune system by triggering reactions against infection and swelling the ranks of the body's killer white blood cells. Consequently, nestlings that hatch into herbal nests cope better with stress and put on weight faster than those without these herbal bouquets. The boost in white blood cell numbers helps the birds survive extreme weather and malnutrition. With all these benefits, nearly twice as many birds from aromatherapy homes return to their nests the following year.

As well as improving the starling's well-being, foliage can act as an insecticide. Many other birds, particularly birds of prey, such as eagles, decorate their nest with sprigs of green leaves. This green garland is naturally laced with high doses of aromatic compounds that deter attacks by insects. The natural insect repellent keeps down infestations of flies, fleas, ticks, and mites in the nest.

The American nuthatch uses sprigs of insecticidal plants to literally sweep its nest clean of parasites. It finishes the job with an even more powerful insect

Above: An **eagle** decorates its nest with aromatic leaves, which deter insect pests.

repellent; it holds a meloe beetle in its bill and smears the beetle's toxic secretions around the nest.

As well as using insects to fumigate their nests, many birds apply insects directly to their plumage. Ants, which release formic acid when attacked, are a favorite choice. Jays and magpies simply lie on an anthill, wings outstretched, and willingly subject themselves to the frenzied chemical attacks of the inhabitants. Other birds, such as rooks, weavers, and starlings, perform a more elaborate ritual. Acting like a contortionist, the bird extends a fanned-out tail to the ground and twists. Then, plucking an ant in its bill, it uses a quivering action to apply the angry insect to the underside of an outstretched wing. As the ant fights back with formic acid, the bird directs the chemical barrage onto its plumage.

At least 250 species of songbird perform this seemingly masochistic ritual, so there is little doubt of its benefits. The formic acid seems to attack the feather mites that otherwise would infest the plumage. The corrosive chemical also destroys fungi, bacteria, and oils, so it probably doubles as a plumage conditioner.

Birds perform the same contortions with a number of pungent artificial substances, including cigarette ends and mothballs. Both contain insecticides; the nicotine that cigarette smokers find so addictive is the tobacco plant's natural defense against insect attack.

A hedgehog will also indulge in ritual contortions when it encounters similar strong-tasting substances. It will lick or chew a strange object until it works up a foamy lather of saliva. Then, throwing its head to the side, it uses repeated flicks of the tongue to transfer the froth to its spines. This self-anointing ritual is easily initiated by a discarded cigarette or a strong-smelling chemical, such as creosote. The frothing and depositing can last 20 minutes or more, until a mess of

Below: A **rook** deliberately subjects itself to an ant attack in order to clean its plumage.

Above: A **hedgehog** anoints its spines with saliva stimulated by chewing on a discarded cigarette.

spittle covers the hedgehog's spiny coat. The purpose of this odd behavior is still unclear but, as the spines are impossible to groom, it is likely that the saliva has some cleansing effect.

In Costa Rica, white-faced monkeys condition their coats by breaking open the fruits of certain citrus plants and rubbing the juice into their fur. They follow the application by vigorous scratching. They also chew leaves and seed pods from various other plants and apply the juice in a similar way. Such topical application is reminiscent of skin ointment and, interestingly, local people use many of the same plants to repel insects or treat skin conditions

When animals take their various treatments, they often appear to enter a trance, as though the medicine were acting as a mind-altering drug. One of the most striking examples is the reaction of the Madagascan brown lemur to millipedes.

*An expression of blissful pleasure suggests the **millipede**'s chemicals might also be narcotic.*

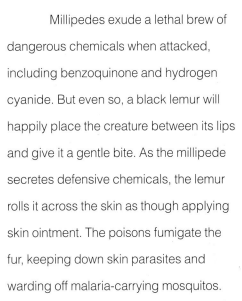

Millipedes exude a lethal brew of dangerous chemicals when attacked, including benzoquinone and hydrogen cyanide. But even so, a black lemur will happily place the creature between its lips and give it a gentle bite. As the millipede secretes defensive chemicals, the lemur rolls it across the skin as though applying skin ointment. The poisons fumigate the fur, keeping down skin parasites and warding off malaria-carrying mosquitos.

As it applies its medication, the lemur drools copiously and its eyes glaze over. An expression of blissful pleasure suggests the millipede's chemicals might also be narcotic. The heavenly state induced by the drug can last 20 minutes or more.

On the other side of the world, South American wedge-capped capuchins use millipedes in much the same way. They also gain protection from mosquitos and appear to enter a similar trance. The event becomes a social occasion, and up to four capuchins may share a millipede, passing it around like a marijuana cigarette. Like human drug-takers, the monkeys are taking a health risk with their habit, because the millipede's secretions are carcinogenic as well as toxic. But the value of keeping down insect pests must outweigh the risks, and the monkeys' obvious pleasure must act as some kind of reward.

Above: **Black lemurs** use millipede secretions to ward off mosquitos — and as a recreational drug!

141

But not all drugs used by animals provide any obvious benefit. Grow a patch of catnip herb in your garden and it will soon become an opium den for local cats. The cats delight in licking or chewing the leaves and flowers and rubbing their cheeks against it. Once under the influence of the herb's heady scent, cats react by staring into space with a blank expression, shaking their head, or rolling over the plant as if in ecstasy. Particularly sensitive cats even seem to hallucinate, for they often paw and chase invisible butterflies or try to catch phantom mice.

*Once a **cat** finds a catnip plant it will return for another fix each day. It is a drug habit that domestic cats appear to share with their larger relatives.*

The reaction to catnip is due to volatile oils known as nepetalactones. Similar chemicals are found in tomcat urine, and this seems to be the key to the apparently pleasurable experience. Some of their reactions resemble those of sexual activity — females roll around in the same way when soliciting attention from tomcats. Three-quarters of all cats react in some way to the plant but the intensity of the reaction varies widely and, remarkably, seems to be genetically inherited. Once a cat finds a catnip plant it will return for another fix each day. It is a drug habit that domestic cats appear to share with their larger relatives.

Above: A **jaguar** chews on a yaje leaf — a plant known to have hallucinogenic properties.

Overleaf: A **cat** opium den. These cats are gathering around a bowl of catnip, a plant that causes them to behave strangely.

In the Amazon, the Tukano Indians believe that jaguars regularly chew a hallucinogenic vine known as yaje. The medicine men, or shamans, of the tribe use the same vine in their rituals. After taking the plant, the shamans believe they are transported into a realm where they can communicate with animal spirits and even become animals. Significantly, one of the most popular incarnations is a jaguar.

When the Indians take yaje they claim it increases visual acuity and enhances sensory awareness. They believe the plant has the same effect on the jaguar and so enhances its hunting powers. Their observations are purely speculative, but it is not impossible that jaguars might have learned that the drug improves senses such as smell or eyesight.

Chacma baboons have been observed eating several plants that cause hallucinogenic effects in humans, and it is likely that they are affected in the same way. To add to the evidence, the way they eat this potentially toxic plant is very different to how they eat normal food, as they are careful to consume only small quantities at a time.

Drug-taking animals may even have been responsible for the Santa Claus legend. The Saint Nicholas we are familiar with was an early 19th century creation, but his roots seem to go back to a far earlier tradition that originated in Siberia among the Sámi people. The nomadic tribes that followed the reindeer herds had a close relationship with the livestock they relied on for food and clothing. Every reindeer was considered to personify the great reindeer spirit, and gifted tribesmen would enter a trance to communicate with the power they believed controlled their herds.

These shamans entered an ecstatic state by eating a hallucinogenic mushroom known as fly agaric. Reindeer travel great distances to feast on these red, white-spotted mushrooms, and the shamans believed the reindeer were

Below: **Reindeer** travel vast distances in search of the hallucinogenic fly agaric mushroom.

affected by the mushrooms in the same way as humans. The shamans sometimes drank the urine of an intoxicated animal to achieve a similar state of ecstasy.

In trance, shamans experience the sensation of leaving the body and travelling to an upper realm — the Christmas images of flying reindeer may have their roots in these visions. The fact that the Sámi wore red, fur-lined costumes and travelled on sleds pulled by reindeer adds weight to the evidence.

As well as being secret drug-takers, many animals show an unhealthy liking for alcohol. Those most prone to drunk and disorderly behavior are fruit-eaters.

Indian elephants…raid illicit stills and go on drunken, hell-raising rampages….

In Africa the fruits of the marula tree provide an irresistible attraction to herds of elephants, so much so that marulas are also known as elephant trees. Once the elephants have vacuumed up the seasonal windfall, the fruit ferments inside them, transforming their stomachs into fermentation vats. Because of their immense size, elephants usually show little outward signs of inebriation. But not all elephants are quite so good at holding their drink. Indian elephants have been known to raid illicit stills and go on drunken, hell-raising rampages through villages. Studies of captive elephants confirm their liking for alcohol; they readily drink vast quantities at 7 percent proof until intoxication takes hold.

Other natural drunks include bees and wasps. Honeybees are partial to the sugary sap that exudes from lime trees. This quickly ferments, so the bees often imbibe an alcoholic cocktail. As can be imagined, the usual beeline back

Guard bees are on the lookout for any **bees** acting disorderly and, like nightclub doormen, they forcibly eject drunkards from the premises.

Opposite: Social drinkers. The **honeybee** enjoys a sip from the fermented sap of the lime tree.

to the nest becomes a drunken meander involving frequent collisions with trees or other obstacles. Even if the bees reach their goal, they often crash-land into the nest or hive and career off course again. Those that successfully alight face another hazard: bee bouncers. Guard bees are on the lookout for any bees acting disorderly and, like nightclub doormen, they forcibly eject drunkards from the premises. Persistent offenders have their limbs bitten off; they literally end up legless.

Birds are also prone to bouts of booziness. Waxwings and robins in North America are sometimes found lying comatose after a day's bingeing on an overripe berries. For lorikeets and cockatoos in Australia, autumn is the season to be merry. Huge flocks of the birds descend on an alcoholic harvest of seasonal fruits and then career around in an uncoordinated manner. The worst cases end up on their backs with their legs in the air.

Our own love affair with alcohol began in prehistory when, like monkeys and apes today, we scoured the forest for ripe, sugary fruit. Because sugary fruit attracts yeast, we would often end up consuming an intoxicating brew of ethanol. As well as causing inebriation, ethanol provides valuable energy. So the two needs became linked, and inebriation became the welcome result of finding nutritious fruit.

Clues to our present appreciation of the bottle are found on the sleepy

Below: Each autumn **lorikeets** in Australia enjoy a fruit cocktail from the leftovers of the harvest. They sometimes end up drunk and disorderly!

Caribbean island of Saint Kitts. In the 17th century, vervet monkeys were brought to Saint Kitts as pets, along with slaves from Africa. Some of the monkeys escaped and set up home on the island. As rum production was the main industry, the monkeys became partial to any fermenting sugar cane lying unharvested in the fields. The slaves

Like human drinkers, the steady drinkers do very well in social groups — they are the life and soul of the party.

soon capitalized on the monkeys' liking for alcohol, and many were recaptured with a mixture of rum and molasses in coconut shells. The monkeys would drink themselves into a stupor and could then be simply picked up. Today their liking for a glass or two continues, and they often raid local bars for a touch of the hard stuff.

A research group on Saint Kitts has studied the vervets to look for any insights they might provide into the origins and nature of human drinking.

Opposite: Happy hour for **vervet monkeys** on the island of Saint Kitts.

Remarkably, the monkeys seem to show the same patterns of alcohol consumption as ourselves. While some monkeys are teetotallers and never touch a drop, others are social drinkers, drinking only with others and in moderation. In line with human drinkers, about 12 percent are steady drinkers, consuming a moderate amount of alcohol each day. Less than 5 percent are binge drinkers — they consume all they can get and readily drink themselves into a coma. The social drinkers prefer their alcohol sweetened, while the steady and binge drinkers prefer alcohol and water at 20 percent concentration — equivalent to a whisky and soda.

Like human drinkers, the steady drinkers do very well in social groups — they are the life and soul of the party. But, in a departure from the human situation, they often become leaders, and they run the troop well. Perhaps this is because vervets are more tolerant of a leader who "monkeys around."

The most intriguing aspect of the study suggests that our liking for alcohol is genetic, and that the tendency to become a moderate drinker or an alcoholic has more to do with genes than social factors.

Such research, as well as recent discoveries concerning animals' use of narcotic or medicinal substances, confirms just how close we are to the other animals that share the planet. We find these similarities strange, if not disturbing; they make us appear less unique, perhaps even a little less human. But we are distinct, although perhaps not in the way our vanity would lead us to believe. We have an odd way of walking; our sexual habits are biologically bizarre; we are strangely defenseless; and we need artificial means to capture and consume much of our food. We also rely on the help of a great number of other animals.

We can only define weird nature by referencing our own behavior. The actions of animals seem weird when they are unfamiliar, but they are equally strange when they remind us of ourselves. Perhaps this is to be expected. The living embodiment of weird nature is closer than we think. It is us.

index

acknowledgments

This book accompanies the BBC series *Weird Nature*. The production was a collaboration of many talents and every member of the team made an immense contribution to developing its themes. Producer Mark Brownlow worked primarily on 'Devious Defenses' and 'Fantastic Feeding'. Producer James Honeyborne was responsible for 'Bizarre Breeding' and 'Puzzling Partners'. Assistant producer Phillip Dalton and researcher Tilly Scott-Wilson worked tirelessly across the series as a whole. Production manager Rachel Kelly, production co-ordinator Margaret Black and production assistant Olivia Hill also provided unfailing production support. Nicholas Pitt worked with many of the animals involved and David Worgan provided practical support. Editor Stuart Napier crafted the resulting miles of film into a seamless production.

 Weird Nature draws on the researches of scientists across the world. Invariably our enquiries were met with invaluable suggestions and guidance and very often the scientists were involved in the filming. I am indebted to everyone who gave their advice so generously.

picture credits

This book is unique in that most of the pictures were taken directly from the programmes. Many of the processes involved were new and the methods required a considerable amount of work and experiment. I would like to give special thanks to Carl Chittenden who worked tirelessly on the computer to extract the maximum quality from the images; Howard Jones who was also involved in some of the digital processing; Cinesite (Europe) who were responsible for producing the high-resolution scans. I would also like to credit all the film cameramen who worked so hard on the series – the pictures are literally a snapshot of their work.

Rod Clarke 12, 13, 14–15, 17, 22, 24, 69 t&b, 79, 86, 94, 97 t&b, 98, 115, 124, 125, 148; John Downer 98; Stephen Downer endpapers, 2, 4 t, 6, 8, 23, 38, 39, 40, 42, 63, 88–9, 92–3, 144–5; Tim MacMillan 10, 26–7, 28, 31, 32–3, 34, 36, 56, 80–1, 82; Peter Nearhos 21, 44–5, 49, 76, 83; Mark Payne-Gill 29, 90, 101, 103, 108, 109, 150; Michael Richards 66, 100, 137; Marcelo Rocha 4 c, 95; Warwick Sloss 4 b, 105, 121, 123, 143; Simon Wagen 99; John Waters 140–1.

BBC Worldwide would also like to thank the following for providing photographs and for permission to reproduce copyright material. While every effort has been made to trace and acknowledge all copyright holders, we would like to apologize should there have been any errors or omissions.
t = top, b = bottom, l = left, r = right, c = centre

Ardea 112 (Pascal Goetgheluck); BBC Natural History Unit Picture Library 16 (Jürgen Freund), 70–1 (Pete Oxford), 73 (Brian Lightfoot), 75 (David Hall), 133 (Nick Gordon) & 149 (William Osborn); Georgette Douwma 72; John Downer 48, 64, 91 l&r, 110, 118, 120, 127, 128–9, 130, 135, 138, 139, 146; Michael & Patricia Fogden 55, 58; NHPA 59 (Daniel Heuclin), 67 (Norbert Wu); Oxford Scientific Films 18–19 (Howard Hall), 47 (Joaquin Gutierrez Acha), 50 (Max Gibbs), 51 (David B. Fleetham), 53 (Karen Gowlett-Holmes), 62 (Lloyd Beesley/Animals Animals), 77 (B.G. Murray Jnr.), 96 br (Rudie Kuiter), 113 (Joe McDonald/Animals Animals), 117 (Rudie Kuiter), 119 (Rudie Kuiter); Mark Payne-Gill: 61, 107 t&b; Seapics.com 84 (Amos Nachoum); Denise Tackett 96 tl, tr, bl.

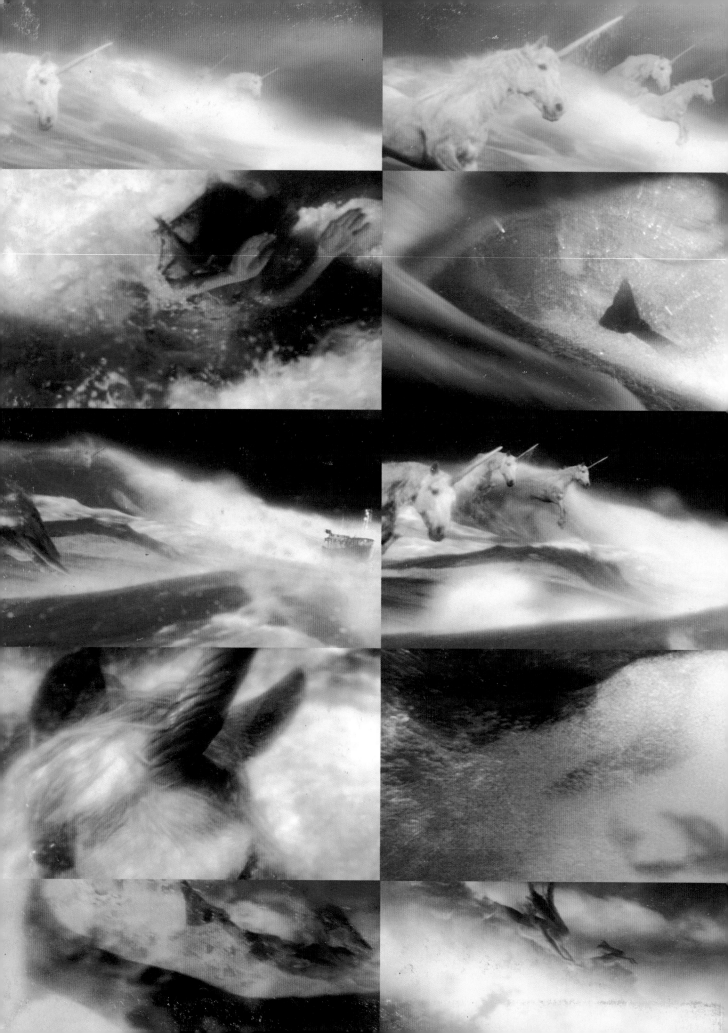